The Executioners

Alicia Cornell Martin

Thank You

I have been blessed with a wonderful family and many special friends who have been God's instruments to teach me truth. Mom and Dad, your instruction gave me a firm foundation, and your example gives me a blueprint for a life that honors God. Missy, you are my very special sister, friend, and sounding board. Grandma Heidi, you are one of the biggest reasons I love writing; your love and encouragement are woven into these pages. To all four of my Grandparents: your two marriages together represented more than a *century's* worth of commitment. Following that legacy, Mom and Dad, your marriage is one of the most spectacular blessings of my life. <u>Thank you</u> for all the hard work.

I could fill pages with the names of additional family members and friends, all of whom provide the kind of support and encouragement that humbles me with gratitude. You have prayed for me, encouraged me, taught me, and provoked me. You have cried with me, laughed with me, and worked along side me. Thank you so very much, and may God pour out his richest blessings upon your lives.

1

Dr. Edward Kline was a trim, distinguished-looking man whose once-dark hair was now almost entirely silver. He sat behind his desk and looked down at the evidence in front of him. It stared back. A single manila folder, perhaps an inch thick. He didn't want to deal with this problem. But it was all his. When he thought about how much power he was able to wield, it made him feel heady. For a moment, it felt good to recognize that he was about to make a decision that held sway over billions of dollars and millions of lives. But the good feeling only lasted a moment. It was too much power, even for him. Too much authority. Ironic. He had worked hard – for decades he had worked hard – because he thought he wanted the power and authority that had come with this job. His mind drifted back to the day he had set his sights on the chair he now occupied…

He had been an undergraduate in a pre-med program. He had known for years that he wanted to be a doctor. Now he was actively pursuing that dream. He wanted to help people, to heal them. When he learned that politics sometimes tied healing hands, it was a rude awakening. That was wrong. So he fashioned his medical training so that he would be in a position not only to help heal people, but also so that, eventually, he would be one of those powerful persons who had authority in the realm of medicine.

* * *

Linda Frederick looked at the tiny cell phone in her hand. Sometimes she wondered how the microphone picked up her voice so well, since it was so far from her mouth.   Enough wondering about useless things.   She dialed the phone number that rang directly to the rich-looking red leather chair in which Dr. Kline now sat.

* * *

When Edward heard his phone ring, he knew it had to be one of two people. He didn't want to talk to either of them, so he let it ring. Five rings. Seven rings. Ten rings. It was Linda. His wife would respect his wishes to be left undisturbed and hang up after the fifth or sixth ring. Now he DID want to talk to her. Judy. The woman who loved him.   The woman who stood by him through medical school, internship, a private practice that often involved babies that wanted to be born in the middle of the night, and a move to a brand new city. The move had been the hardest on her.   She really did not want to leave the peaceful mountains of New York and start life over in a new location farther down the eastern seaboard. But she loved him, and so she had. As the phone continued to ring on his desk, Edward had one of those moments of incredibly deep appreciation for his wife. Two minutes ago, he had not wanted her to call because he knew that the moment she heard his voice, she would know that something was troubling him greatly. Now, he longed to hear her voice, and to know that the woman he loved grieved over his troubled heart, even though she had no idea what the trouble was.

By the time the phone hit the eleventh ring, Dr. Edward Kline knew what he had to do.   It was an incredibly huge decision; one which would certainly change the course of the rest of his life. It was the kind of decision that most people would never make in less than 24 hours. But as he recalled why he had wanted this chair, and as he remembered Judy, the choice had become clear.

Literally and figuratively, the phone would never stop ringing until he answered it.

"Hello Linda."

"Dr. Kline, how did you know it was me?"

"I could say it was a lucky guess, but that would not be true. The truth is, as soon as I saw this manila folder, I knew you'd be calling. I still haven't figured out how you would know when I received it, but obviously, you have a way of seeing inside this office. How much are you willing to pay me to 're-interpret' this information?"

For a moment, Linda Frederick stumbled. Dr. Edward Kline was known to be an honest, honorable man. Yes, she had called him because the information in the folder needed to be suppressed for a few weeks. But her whole strategy for this phone call had been one of temporary suppression; she had never considered that Kline would take a long-term pay-off. She was not prepared in any way for this possibility.

"What do you mean, 're-interpret?'" Linda was trying to buy time while her mind scrambled.

"You know exactly what I mean. How much?"

The dead silence on the phone stretched on for three seconds, and Edward Kline smiled for the first time in five hours. "It's a simple question, Linda. I'd say you have access to at least $350 million – don't you?" Three more seconds of dead silence. Edward Kline was learning exactly what he wanted to know. "The next time you call, be prepared to offer $500 million." And then Edward Kline hung up.

* * *

"Susan, do you know where my soft, blue short-sleeved shirt is?"

"Try the second drawer from the top in your tall dresser." Sure enough, on the left-hand side, third from the bottom, was the shirt Richard wanted. He shook his head. He knew Susan was not feeling very well today; her last chemo treatment had been four days ago, and it usually took her at least five days to get past the worst of the effects. She wanted to walk on the beach, though, and he was thrilled to take her. He was so thankful that she wanted to get out, and after one of their walks, she almost always looked a little better. Deep down in his heart, though, Richard could see that the doctor had spoken the truth: Susan would probably not live another year.

* * *

"What's wrong?"

Edward Kline was frustrated and relieved at the same time. Judy had asked the question almost as soon as he had walked through the door. He didn't want to talk about the manila folder at work; legally, he shouldn't, and this frustrated him. But he was also relieved – so very relieved – that his wife loved him and cared enough to know and to ask. He gave her an enigmatic smile and, to his surprise, his eyes got a little moist. He took a deep breath and looked at his bride of 46 years. "I really can't say. But thank you for seeing it." And then he hugged her.

Judith Kline was wise enough to know that, for now, just a hug was what her husband needed the most. Then they had a nice dinner in the sunroom and watched as the sun set.

Two hundred miles east of the Kline home, Richard and Susan Serrap watched the same sun set as they walked the beach at Point Lynne.

2

Dr. Kline arrived at his office the next day and retrieved the manila folder from his office safe. He read the whole file again, knowing that he had to become familiar with every aspect of it. His charade on the phone last evening with Representative Frederick would buy him little time. She wasn't the only one who wanted the information in it suppressed. In fact, there was a whole lot more money that would work to suppress this file than there was money to help it be birthed into the realm of public knowledge. Edward smiled at the analogy that came to him, not because it was humorous, but because it was so appropriate: this file was sort of like one of the many babies he had spent his life learning to save. Tiny, fragile lives who fought to live, who *should* live, and yet the odds were so stacked against them. Edward's hands had saved many such lives. Sometimes, he had literally fought for hours on end, pouring every ounce of himself into the task, to save them. The contents of this folder were not living in the literal sense of the word, but there were many who would try to kill them. And the odds were against this little folder. If Edward could save this one…it would be the hardest "baby" he ever helped to live and, in a way, the most important. And live it should. Live it *must*. But how?

"For such a time as this…" The words echoed through his mind. "Perhaps you have come into the kingdom for such a time as this…" Those words were once spoken to a young queen whose royal position had been used to save the lives of her kindred. The fact that she was the queen was what had allowed her to save their lives. Dr. Edward Kline, pacing around his office, looked at the leather chair behind his powerful desk. He was not royalty, but perhaps he had come into his powerful position in the medical community "for such a time as this." After all, his Chair was probably the closest thing to the Royal Throne in the kingdom of medicine. The queen in his memory had used her position to save the lives of people unjustly sentenced to die. He was in a very similar situation. It was his responsibility to use the power of his position to save the lives of people who had unjustly been sentenced to die. Here, in the 21st century, it was not an edict which attempted to steal lives; it was willful ignorance. The knowledge in this folder *must* live. But how?

Edward Kline's mind went into "emergency-doctor-mode." It was a honed gift that he had. When a tiny, high-risk, unstable patient lay before him, the vision of his mind would narrow onto that one little person. Within a minute, sometimes seconds, he would have decided upon a preliminary route of treatment. And then, using every resource at his disposal, especially decades of experience, he would set to work.

What was the most "life-threatening" complication regarding this manila folder? How should that be dealt with? What other "health" complications did he know of? Were any of them going to get worse if not dealt with immediately? Were there some potential problems which might arise within the next few hours, or days? Were there any discernable clues to suggest major "health risks" other than the ones he was immediately aware of?

Representative Frederick and her constituency were probably the biggest risk that the information in the folder faced. How could he deal with her, and the power and money behind her? He knew that they would stop at nothing less than burying the information. He had asked for a bribe last night because he knew that it would throw her off her guard, and buy him time. Time. That was his most precious commodity right now. How could he get more of it? He could ask for more money, no matter what figure Linda came back with. But if he came across as too greedy, could that get him killed? The thought struck him with force. Just how much life-threatening danger was he in?

Edward had sort of considered and then dismissed the thought that there were powerful people out there with enough money to have him killed over this knowledge. But no. Too many researchers had contributed to this folder. They could kill him, but the information *would* come out. Wouldn't it? If Edward were gone, who would be sitting at his desk? He was not comfortable with the answer. The man was a good doctor, but he might be inclined to interpret some of the data in Frederick's favor. No, the information could not be concealed forever, but if the manner and timing of its discovery were carefully controlled, it could be glossed over for years. And tens of thousands, maybe millions, of lives would pay the price.

Suddenly, one of the contingencies Edward had to plan for was his own death. As the reality of this sunk in, it was incredibly difficult for him to temporarily lay aside the deep emotions of how much he loved his wife.

* * *

"Four hundred and fifty million. It's the best I can do."

"And what exactly do you think it will buy you?"

"In your own words, doctor, it will buy us a 're-interpretation' of the evidence."

In his own words. Did she TAPE that call?!? No!!! But Edward's voice betrayed none of the panic he suddenly felt. "I said five hundred, and I meant it." Click.

He had planned to hang up suddenly like that, having decided that the time he might buy was worth the risk. But now he was extra grateful to be able to hang up, because he wasn't sure if he could have thought straight wondering about tape recordings. His mind was trying to re-play that first phone call, and figure out what was on the tape, if she had recorded it. Had she spoken his name, and he acknowledged? He couldn't remember.

Edward took a few minutes, and forced himself to calm down. Yes, she could have taped that first call, and maybe the second one, too. But he could think of no way of finding out for sure, and there was no use worrying about it. He would try to plan for the contingency. That was the best he could do. One thing, at least, he had on his side: Linda had not expected him to ask for a bribe, so the odds were that the first call was not taped. Edward had never liked the idea of relying on "odds." Nevertheless, in his profession, when he spoke with patients and their families, he referred to them a lot. From a logical perspective, he found some comfort in knowing that, when the odds were in your favor, it was a good thing.

* * *

Representative Linda Frederick was confused, but in a pensive, calculating way. What kind of game was Dr. Kline playing with her? He had a reputation that flew in the face of the way he was acting right now. Had the honorable doctor put on an act for years, just waiting for the opportunity to make a lot of quick, easy money? Or was he toying with her? She had never expected him to ask for a bribe, but when he did, she was pleased. It was as though he had offered a simple solution to a complex problem. She had deluded herself into believing that Edward Kline recognized the insurmountable odds against publicizing this now-documented knowledge: he would never be able to fight all the power and the money, so why not join the other perspective, avoid a losing battle, and make some money? Linda Frederick lived her life that way, and so she was able to convince herself, in the beginning, that Edward Kline might do exactly that.

It had taken Linda a little work to convince the money brokers that Edward Kline really could be bribed. They had agreed to $450 million. Now a nagging doubt clawed at Linda's mind. Not even she could swallow the idea that Edward Kline, with an impeccably spotless reputation for honor that spanned decades, was going to hold out for a $500 million bribe, when she had just offered him 450. If he was going to take the money, he would have taken the 450. She'd been a fool! Worse, she had to go back to the money brokers and tell them that the problem was not solved.

*   *   *

"How do you feel, sweetheart?"

"I've felt better."

Richard knew that. Susan looked tired, drawn, fragile. Was it really worth it? The chemotherapy, and now radiation, gave her a small chance of beating this thing. But it was such a small chance. And it was a lot of suffering to pay for that small chance. Richard held Susan's hand tenderly, stroking the back of it with his index finger. He *loved* her! He did not want to lose her; could not bear the idea. But he loved her so much that seeing her suffer like this was worse than the idea of losing her.

There was so much they had never done. Places they never saw. Children they never had. Richard fought to make sure a second tear did not follow the first down his cheek. Children. For years they had talked about beginning a family, and then for years they had tried. They learned too late that mistakes from their past had made it impossible for Susan to conceive. It had been heartbreaking, especially for Susan, who blamed herself. But forgiveness had brought healing, and when they were able to adopt Jacob, the love that they poured into him began to cover over the hurtful scars from the past. They had wanted to adopt at least one more child, but that had never happened.

Now, as Susan lay dying, Richard was more thankful than ever for Jake. He was a blessing to his mother, and to him. If God took Susan home, at least Richard would still have Jake. He was overseas right now, with his parents' blessing, but he kept in constant contact.

When Dr. Edward Kline saw the Mustang pulled to the side of the road with the driver's door open, he was suspicious. The last several days, it seemed, he was suspicious of a lot. He knew he was trying to be careful, but he also knew his actions were bordering on irrationality. He slowed down, still unsure about stopping. In this age of cell phones, help would certainly be on the way if it was needed. But then his eye caught sight of a man who had apparently walked around to the brush on the passenger's side of the car. He was bent over, on his knees, retching. And then he sunk to the ground, as if devoid of strength.

Edward did not dismiss his suspicions, but neither could the doctor in him ignore an obviously ill man. His left hand touched a pocket in his jacket, as if to be reassured of certain contents. Edward slowed to a stop near the sick man, and got out of his car. Having noted that one of the man's hands was not visible, Edward removed the envelope from his pocket, dropped it, and shoved it into the brush along the road with his foot as he walked around his car. As he approached the prone figure, he identified himself as a doctor. "Thank goodness," came a barely audible response.

"Do you need me to call for help?"

"I don't know, Doc. I stopped for a hamburger a few miles back, and I was fine then. I don't think it agreed with me. Actually, I feel much better now, just weak and tired."

"May I examine you and ask a few questions?"

"Of course. It's a blessing to have a doctor who stopped."

Thirty minutes later, Edward was getting back into his car and pulling away. The man seemed to have had a case of food poisoning. He was feeling much better, although a little weak. Edward was fairly certain that the man would be fine now. He had assured Edward that he only had to drive five more minutes before he was home, and that his wife would be there.

As the doctor drove, he thought about how thankful he was that his initial misgivings about stopping to help the man had been unfounded. The envelope! He had forgotten to retrieve the envelope from the side of the road before leaving. He would turn around up ahead, and go back and retrieve it. He was thinking about retrieving the envelope when his eye caught the dump truck barreling towards him – on his side of the road. Edward did his best to avoid the truck, but it was too late. He asked God to watch over Judy.

* * *

Another car stopped at the scene of the horrible accident. A man in dress slacks jogged towards the wreckage, apparently seeking to help if he could. One of the people on the scene who had previously stopped caught the man's eye and just shook his head. The jogger understood, closed his eyes, and grimaced. And then he sank down on one knee, holding his head in his left hand. If anyone had been watching him, he seemed to be just another stranger who was emotionally disturbed by this gruesome reminder of death. He stood up again soon after, and walked back to his car, head down. No one had noticed that, as he squatted beside the car, his right hand had reached under the bumper.

* * *

Judith Kline had been watching out the window when the squad car pulled up. She had expected Ed home more than an hour ago, and he had not called. One look at the officer's face, and she knew that she would have to lean on the strength of her Lord, because her own would not be sufficient to face the coming days.

"I'm so sorry ma'am…" And then the rest of that evening was like fragments of a nightmare. Some parts she remembered vividly, and other parts were so nebulous that, later on, she could not recall them. "A stolen dump truck… The driver fled the scene… We're looking for him… Is there another family member living nearby whom we can call to help you?" A kind woman put her arm around Judith, and did not say a word. Judith wondered later if that woman had any idea how much comfort that had been.

* * *

"What do you mean, he's *dead*?!?" The voice was Linda Frederick's.

"A stolen dump truck lost control and crashed into the driver's side of the car on his way home from work. He died instantly."

"*Stolen*? By whom?"

"Yes. Stolen. The police have a suspect, but they have not found him yet. He fled the scene of the accident. Apparently some kid took the truck for kicks, and couldn't drive it too well. It seems he got scared and ran when he realized that his recklessness had actually killed somebody."

"I don't believe that story."

"*You* don't have to."

* * *

The funeral of Doctor Edward Kline, husband, father, grandfather and friend, was attended by more than

500 people who loved him dearly. It was also attended by two plainclothes police officers. They were the two who had come to the door of the Kline home a week ago, and Judith had graciously given them permission to attend. She thought that she might know why they wanted to come, but right now, that was more than she could handle. She needed to grieve first.

<h1 style="text-align:center">4</h1>

"I'm just not convinced that it was an accident."

"Why?"

"Because if the driver did not have the intent to hit that car, I think he would have ditched the truck – there *was* room to do so."

Richard's supervisor, Ken, squinted his eyes a little, so as to look at the scene of the death in his mind's eye. "You might have a point. But, Richard, I've been behind the wheel of an out-of-control vehicle before. It's also very possible that the kid tried to ditch the truck and simply couldn't. He was fighting literally tons of mass, and we have no idea whether he was drunk, or high, or what. Besides, what possible motive would a teenage boy have for killing a respected medical doctor?"

Richard had to accede to Ken's logic, and did so with a nod. Still, something suspicious gnawed at his gut. "Christopher Mills was a rebellious young man, and no one was really surprised by the idea that he stole the dump truck from the county road maintenance department. And, I am convinced that the witness had a decent view of the crime to positively identify him as the thief. But what did he do with the truck for five hours between the time he stole it and the accident? I just wish we could find him."

"It's a good question, and I want you to find him just as much as you want to find him. Keep looking."

Richard went back to his desk, took out the case file, and laid it all out on his desk. He only sorted through it for a few moments. He knew it all; there was nothing new here. He had to look somewhere else. Putting on a light jacket, he went out to the evidence yard and began looking over the truck and the car. Experts who were more trained than he in knowing what to look for had already done this, and he had read their report. He was not going to find anything they had missed, but it was good to have an accurate picture in his head of what they looked like. He drove to the scene of the death. He drove the route from Dr. Kline's direction, and then from the dump truck's direction.

Richard parked his car at the scene, and got out. Christopher Mills fled the scene on foot – he thought. But no one had witnessed the actual crash. This little country road that led to the Klines' home was not overly-traveled. It did not even have a yellow line painted on it. He thought back to his comment in Ken's office about ditching the truck. Had there really been time? Yes, he thought so – if the young driver indeed had the ability to do so, and that was not a foregone conclusion. Ken was right – a good driver might have trouble ditching an out-of-control car, to say nothing of a young driver behind the wheel of a fifteen-ton dump truck. And maybe Mills had gotten high or drunk in those five missing hours.

Richard pulled his mind back to fleeing the scene. Everyone had assumed that Mills had fled on foot. Had he? Could he have hitched a ride with someone? Doubtful. In order to do that, he would have had to encounter a car driving towards the accident as he fled it – which would mean that the car would bring him right back towards the accident. The car then would have surely stopped to lend aid, placing Mills right back at the scene. Could he have had someone or something waiting? Not unless it was pre-meditated, and even then, it would be nearly impossible to plan the exact location of the collision; the doctor did not always come home at the same time of day. Richard could not imagine Mills pre-meditating something like this, and started to dismiss the idea. But then he paused. His years of detective experience had taught him that the most easily-dismissed idea sometimes bears looking into – if only for the sake of the rabbit trails that his mind could find by following it.

Richard began talking out loud to himself. "Let's say the kid did pre-meditate it. How in the world could he make sure that he was on the road at the same time as the doctor, and how could he make sure that it was the doctor that he hit, and not someone else?" Richard had been walking around, looking at the ground. Now he stopped, looked up, and began looking around. He was in the middle of a clearing. A bend in the road decreased visibility for drivers, but if someone was standing on that hill over there, they would have a clear view of the car approaching from Kline's direction a couple of minutes before it got to the clearing. If the lookout signaled the truck to come, would the truck have time to reach 40-50 mph by the time *it* reached the clearing? Probably, depending upon exactly where the truck was parked.

Richard stopped and shook his head. The detective in him was nearly certain that Christopher Mills would never have committed such an elaborate crime. He stole the truck for fun, or for drug money, maybe, and then lost control of it while driving wildly on a back road. Fear and/or guilt motivated him to flee the scene. They would find him soon, and the charge would be involuntary manslaughter. That's where the evidence all pointed, and if they had found him already, that is where they would currently stand on the path to justice. But where *was* he?

* * *

Linda was pleased to be informed that Dr. Carl Huxley had officially been seated in Dr. Edward Kline's vacated Chair. The manila folder was now in the hands of someone whom she trusted. Still, she had to contact him. For a moment, she resented the task, because it made her feel like a lap dog. She was *Representative Frederick* – a woman of importance and power. She should have underlings who did these kinds of jobs for her, rather than her having to do them, as though she were the underling for someone else. But the resentment faded a bit when she looked at the situation through their eyes. They had the money, but they did not have the power to pull their own strings. They *couldn't* speak out, because as soon as people realized who was doing the talking, the game would be blown. They had to present themselves as the innocent good guys; the moment *they* called Dr. Huxley, even the most naive reporter would recognize the financial motive for the call. No. The only way they could maintain the "good guy" image was to be squeaky-clean ignorant of the truth – in the public's eye, at least. The fact that they were paying her to make the phone call, Linda realized, was all the evidence any court of law would need to prove that they were not at all ignorant of the truth.

Linda drummed her fingernails on the desk beside her phone. She was not their lap dog. She was one of them, and just as powerful as they were. In fact, perhaps she was amongst their most powerful members. She was one of the few who *could* interact with powerful people in important chairs so as to shape public policy in their favor. They *needed* her – without her, they would not have a phone line to the throne room of medical law. Linda smiled. She was feeling much better about the phone call she had to make. It was a privilege.

* * *

Richard, having dismissed the notion that Christopher Mills would pre-meditate the murder of Dr. Kline, began walking back towards his car. They needed to find the young man who, undoubtedly, was afraid to be prosecuted for the death he had caused. That's why he was hiding. But as long as it *was* an accident, Richard resolved to do everything he could to help the young man. He would have to pay for the pain and death he had caused, and he would have to live with a measure of regret for the rest of his life, but Richard did not want to see the young man's life ruined over the ordeal, either.

Richard was becoming comfortable with this whole train of thought when an object caught his eye. Squatting down, he looked at a little black piece of plastic, examining it without touching it. His heart began to beat faster, and his brain traveled at break-neck speed back to the rabbit trails involving pre-mediated murder. Like a computer running a thousand different guesses per second, trying to figure out what a password might be, Richard's mind began tracking a whole new set of possibilities as to where Christopher Mills might be. A few seconds ago, his concern was that the young man's life would be maimed and crippled by a foolish decision. Now his concern was that the young man's life might be forfeited altogether.

5

"Ken. I need the forensics team back out to the scene of the Kline accident. I found a tracking device in the field near the collision. I'm almost sure it was thrown over here upon impact. I think it's time to change the primary hypothesis from accident to homicide."

"They'll be there this afternoon."

* * *

"Dr. Huxley, Representative Linda Frederick is on Line 1."

"Thank you. ...Representative Frederick, it's a pleasure to hear from you."

"Congratulations, Dr. Huxley. I was pleased to hear of your promotion."

"I'm sure you were." Dr. Carl Huxley then stopped talking. He knew of Linda Frederick, and he thought he knew why she was calling. But this chair had only been his for a week, and he was not going to lose it through stupidity. He would let her talk.

Linda was not caught off-guard. She knew that she was going to have to take the lead in this first dance. After that, her skill would lay in allowing the good doctor to think he was leading. "How many big problems did you inherit with that chair?" The tone was meant to sound sympathetic.

Dr. Huxley smiled, since she could not see him. He would tell her "one," but the honest answer was "two" – because he had inherited her phone calls, too. "There was only one major research issue that was still unresolved when Edward died so tragically. I've read through the initial reports. The evidence appears to be cut-and-dry at first glance."

Linda did not miss the comment about Kline's death, but let it pass. "Appears? At *first* glance? Sounds like things may not be exactly as they appear, and that the second glance may be worth taking." Now it was time for *her* to stop talking, and see where Huxley chose to lead. There was a moment of silence.

"You're right. I think it's important that we don't allow religion to dictate the conclusions of medical research. Of course, if science learns how to prevent some diseases, then it would be unethical to cover that up." He paused.

"Of course." Linda knew she had to acknowledge this; she was talking to a medical doctor.

"However, even those on the conservative side of the fence are the first to admit that it's wrong to destroy someone's whole life in order to prevent disease."

"You're absolutely right, Dr. Huxley! That's a brilliant way to state the case!" The admiration in Linda's voice as she spoke was not faked. She was really impressed with Huxley's wording. He had thought out the situation and found an ingenious way to turn their opponents' words against them – or so it seemed to her. She was also relieved and assured that Huxley was on their side. He would lead the dance right where she wanted to go. Only the technicalities had to be ironed out. "When you finish analyzing that second glance, why don't you let me know? Perhaps I can help get the necessary information into the hands of the right people."

"I would appreciate that. You'll hear from me soon. Have a good afternoon."

"You, too." Click.

Huxley smiled again. The admiration in Linda's voice fed his ego. He had sought to impress her, and had succeeded.   This would be a pleasing and lucrative friendship. The benefits of his new job were at least as spectacular as he had hoped. He was a king in the world of medicine. He moved amongst the rich and powerful. A beautiful Representative needed him – or so he thought. And the financial benefits from this phone call, while perhaps slow in maturing, would probably be incredibly significant.

* * *

"Did the team find anything else?"

"No."   There was a bit of disappointment in Richard's voice, but not too much: he was truly thankful for what he *had* found. "But that doesn't bother me as much as the fact that Mills is still missing. Before, I figured he was just scared, meaning that he would turn up sooner or later. But now... if this was a homicide, he may have been used... and then disposed of."

Ken nodded. "I've been thinking the same. But consider: if the intention is to make it look like an accident, and Mills was used, having him permanently disappear is not the ideal scenario. The body should have shown up shortly after the accident with a suicide note or something."

"I agree. But if I hadn't found the tracking device, this would still be an accident. In fact, as I was leaving the scene before I found it, I had myself pretty much convinced that it *was* an accident. If it is a homicide, and the killers resorted to a tracking device, then I am sure they tried to retrieve the device and failed. If it wasn't for this mistake, this death would have been ruled an accident even if Mills never showed up. It would have been assumed that he ran away and began a new life somewhere else."

"Your theory seems pretty ironic – they failed to retrieve a tracking device? By definition, it's an easy thing to find."

"Unless it's broken. The lab isn't finished with it yet, but I'll wager it doesn't work any more. Someone could have looked for it on the car, but if it was not transmitting, the odds of finding it where it lay in that tall grass would be a million-to-one."

"Maybe I should let you buy my next lottery ticket."

Richard smiled as he got up to leave the office. "Don't waste your money. Those things are just a tax on people who don't understand probability."

6

Probability. Richard had joked with Ken about it a few short minutes ago. But now as he started home, he couldn't help but dwell on the odds of Susan living another year. They were so slim. She was fading away before his eyes, and he felt so helpless.

"How do you feel?" Richard directed the question to his wife as he kissed her on the forehead. The smile on her face contrasted sharply with the gaunt, pale skin.

"OK, Rich. The nausea was pretty much gone today, and that's a blessing. I just wish... No. I won't go there. I'm thankful that you're home, that I don't have to go through this alone. And I am confident that God can take even this and work it for good."

Richard sat down beside Susan, and put his arm around her – the way he knew she liked. She leaned into him, and he stroked her thin hair. "I know you're right, Sue. But I..." His voice broke. "...I hate to see you suffer so, but I can't stand the thought of losing you, either. I'm so proud of you for not giving up."

Susan Serrap was a woman of strong faith, and her Lord had made her stronger throughout this war with breast cancer. But sometimes, like now, it was a blessing beyond comprehension to be able to lean on her husband and let him help carry the grief. She reached out her hand, and just held his. Richard understood this request, and in a broken, but strong and thankful voice, began to pray for the strength and discernment that would be needed in the coming days.

*   *   *

"You were right – it doesn't transmit anymore." The lab technician was briefing Richard. "That's the good news, since it would have disappeared from the accident scene otherwise."

"And the bad news...?"

"I couldn't pull any information out of it or off of it that might indicate who was using it."

"When it worked, how far could it transmit?"

"A mile or so."

"That's not far..." Richard's brain immediately began running scenarios of how the device could have been used. The technician recognized the glazed look in his colleague's eyes. He had already told Richard as much as he knew, and at this point it was better to leave the detective alone. He smiled, glad that he seemed to have provided some needed information, and quietly left the room.

Richard was imagining the accident scene. "...a mile or so..." What's the use? To be alerted that the car is coming near. That little hill in the distance. Someone sat there, waiting for the signal to show up. That meant Dr. Kline was close. The lookout calls the dump truck and tells him to start their way. The dump truck gathers speed. The lookout can easily tell, with a pair of binoculars, that the car is, indeed, Dr. Kline. He can also watch and make sure that there is not another car close enough to Dr. Kline to witness the accident. If such a car were there, the dump truck is told to abort, slows down, and passes the two cars as normal traffic. Then they wait for the next day, and another opportunity.

How many days did the murderers wait? Only one, assuming they only had the one dump truck; it had been stolen earlier that same day. Where was the truck waiting? If Richard's theory was correct, the truck had to have been parked pretty close to the accident scene. He had to climb up on that hill and look for any evidence that someone had waited up there. He returned to the accident scene and vicinity...

* * *

"Ken, I've got a lead." Ken looked up. Richard had a map, which he laid out in front of Ken. There were three "X"s on it. Richard sketched his theory, pointing to the various locations on the map. "I couldn't find any evidence that someone spent time on the top of that hill, but that doesn't concern me. They were careful. But back over here is a little pull-off, just the right distance from the field for the dump truck to wait. There are tracks from a big vehicle there and, guess what? They match our truck."

"Good work. Now the people. We're going to have to talk to the widow, the co-workers, etc. We've *got* to find Mills."

Richard gave his superior a polite nod and left the office.

* * *

Richard looked up from his desk to see Ken with a disappointed look on his face. "We found Mills."

Down in the autopsy lab, Richard conversed with the coroner. "Does it look like he killed himself?"

"Yes."

"But it's possible that he didn't – and I need you to find the evidence for the truth."

"I'll do my best, Richard."

"I know."

* * *

"Richard?" It was the coroner's voice, and Richard's hope rose at the tone in his voice. "I'm confident that someone else's hand was holding Mill's hand when the trigger was pulled. I'm going to keep looking, but I wanted to give you that right away."

"Thanks, Jim. Good work."

* * *

As Richard's finger touched the doorbell, he wished desperately that Susan could be with him. He did not like having to talk to widows of murder victims. Many knew that, of all the detectives in the office, Richard was the best man for this job; his kind, compassionate heart really grieved for the women, and they could see that. All Richard knew was that Susan would be a tremendous comfort to this woman, and that this conversation would be easier if she were beside him. Not that Susan ever came with him to work-related calls; but he thought of her often at such times.

Richard had called Mrs. Kline earlier, giving his name, and explaining that he was the detective in charge of investigating her husband's death. He told her that he had found some more information that he needed to share with her. He wanted to ask her a few questions, if she was up to it. She had graciously agreed.

"Mrs. Kline?" The older woman who answered the door reminded Richard very much of his own mother. She smiled kindly at Richard and the officer who accompanied him.

"Thank you for calling earlier. Won't you please come in? My son is here with me, and we can sit in the living room." Judith Kline led them to a comfortable, inviting living room. Richard noted that this was probably the room where the Doctor and his wife had spent many evenings together, and it was not hard for him to guess which chair had been the Doctor's. Out of respect, he carefully avoided sitting in it.

Mrs. Kline brought a coffee pot and set it on a little table, where she had already placed mugs, sugar, and milk. At her invitation, Richard made himself a cup. While he did so, the son came in, and she introduced him. Then they all sat down. Mrs. Kline sat quietly, waiting for Richard to begin.

Richard marveled at how composed she was. The pain was evident in her face, but clearly she had determined that she was going to try not to cry. Richard began by explaining that the young man who had stolen the dump truck had been found. When he shared that the boy had been found dead, along with a suicide note, the grief in the widow's face heightened. She did not say anything, but Richard understood the expression on her face and her body language: there was no animosity in this woman towards the boy; she had already forgiven him. The fact that he was dead, apparently by his own hand, brought her a sadness that added itself to the grief she already bore.

Richard had made the decision that he would tell Mrs. Kline that her husband's death had been re-labeled as homicide. She and the son in the room would be the only ones for now, though. He needed some direction, and he hoped that this woman could give it to him. Slowly, he broached the subject.

"In spite of the suicide note, though, we think that Christopher Mills may not be responsible for your husband's death." Richard expected the look of confusion that followed, and continued on. "It's possible that someone planned your husband's death, and that they used the boy to steal the truck, and then killed him, too." Richard stopped for a moment then. The Klines needed time to digest this. He had learned that, at moments like this, it was almost as if the death of the loved one happened all over again. In their minds, at least, it did have to happen all over again, because the explanation that they had been learning to live with and accept was no longer valid. It meant going back to the moment of learning about the death, and starting over. And that was hard. And painful.

After a few moments, neither of the family members jumped in with a question, so Richard continued. "I know this is hard, and your minds are probably reeling. I've heard nothing but kind and admirable things about Doctor Kline. Can either of you give me a clue as to who might be so desperate as to do such a thing?"

Both the widow and the son were still wading through the implications of Richard's words. He patiently waited.

Neither Judith Kline nor her son, Robert, had remotely considered the possibility that someone would murder their husband and father. He was an honorable, respected, well-loved man in his family, his Church, and his community. They knew that he worked with some people who disagreed with what the Doctor knew to be true. That was, by far, the greatest source of tension in his life. But that one of those people might...kill...the word was so horrible. And yet, here sat a police detective who apparently had enough evidence to make him seriously consider that possibility...

"What makes you think that this was intentional?" Judith voiced the question that Robert knew she wanted to ask for herself. Richard sketched the evidence up to this point, emphasizing that there was still very little to work with.

Again, silence reigned in the room. Richard knew that these grieving people were asking themselves the same question he had asked: who? Finally Mrs. Kline looked at him. "Detective Serrap, I just don't know. Ed just did not have 'enemies.' All I know to do is to encourage you to look at what he was working on at the office. The last few days before he died, something had been really bothering him. I was used to that. Sometimes he would talk to me about it, sometimes not. He did not give me any details this time, but I have a strong suspicion that he was about to make another politically incorrect decision.

"Ed felt so honored to be able to sit in that Chair and make decisions about the laws that govern the medical community, and which direction research went. Those decisions always had a single focus: do what was best for every single patient. With politics and culture the way they are today, that meant that Ed's decisions were not always very popular. When he refused to condone fetal stem cell research, he thought that was going to be the end of his tenure. But then truth prevailed, and so did Ed's privilege to remain where he was.

"When he came home a couple of weeks ago, I was reminded of that battle. I thought maybe it had cropped up again. If it had, Ed did not say. But whatever was bothering him, I'm pretty sure it was something like that. My husband was a strong man, who knew the right thing to do, and who would determine to do it. But as I'm sure you know, that's not always easy. Ed would often know what he had to do, but he would wrestle over exactly how to do it, so that the fewest number of people got hurt."

Everyone in the room was silent for a moment, out of respect for what Mrs. Kline had said. Once it was obvious that she had finished, Robert spoke up. "Detective, I can't tell you much more than what my Mom just said. Dad called me a couple of weeks ago – he calls often, so that wasn't unusual. But before we hung up, he did ask me to pray for an upcoming decision that he had to make. It's sort of funny for me, hearing Mom say what she did just now, because of the way Dad worded his prayer request. He said that he knew what he had to do, but he needed lots of wisdom to know how best to do it." Robert's eyes were a little moist by the time he finished talking, and he looked over at his Mom and smiled.

Once the emotion of the moment had passed, Richard stepped into it. "Did your Dad make such requests often?"

"We shared prayer requests with each other very regularly. He often mentioned work, but this one stood out as especially significant – not so much because of what he said, but because of the way he said it. I could tell it was weighing heavily on his heart."

Richard had now learned what he had expected to learn. He was sort of pleased that they had said what he had anticipated. But at the same time, he was not much closer to a murderer with a motive. He requested that they please keep this new information confidential, which they understood. He left them his card with his phone number, and politely excused himself.

*   *   *

"We're in trouble. The death has been reclassified as a homicide."

"We're not in trouble. They don't have enough evidence to convict anybody, and they won't get any more as long as we don't do something stupid. Good grief, Serrap doesn't even have a suspect!"

"And if he finds one...?"

"Who's the best suspect he could find? Someone who doesn't know any names. There are so many layers of caution between us and this incident that the law can't come within ten miles of us. Go to sleep."

The phone hung up, but the worried female caller found it impossible to fall asleep without the assistance of a drug.

7

At a fancy restaurant, Representative Frederick sat across the table from Dr. Huxley. "Are you all settled into your new job?"

"Pretty much. I've been groomed for that job for a while. I just never would have guessed that the timing would be so perfect. But I'm real busy."

"That's good. You're so busy that it will probably be several weeks, at least, before you can finish that 'second glance.'"

"I'm not *that* busy. And if I pretended to be, someone would start asking questions."

"Somebody's already asking questions. It will be in your best interest to get so busy that you don't have time for his questions, or to look at that folder for a while. You might even consider misplacing it – temporarily, of course."

Carl Huxley understood what he was being asked, but much of what he had just heard was new to him. The expression of confusion on his face asked Frederick to give him more details. She volunteered none.

"Who's asking questions? And about what? Clue me in, Linda, if there's something I need to know here."

"Carl, I'm not saying that there's anything you need to know. I'm just sharing a bit of helpful information with you, to help you protect our interests. You implied the other day that that was what you wanted to do. Have you changed your mind?"

If Huxley had been wise, he would have said "yes" at that point and left. He knew he was trapped. This was the point of no return. If he acted on Linda's words, he was "in." "In" – where the big money was, and the powerful nameless princes moved certain aspects of the medical community like pawns on a chessboard. He wanted "in." Besides, it corresponded with his worldview. It wasn't like he needed to hop the fence in order to agree with what they wanted. He had chosen long ago to define ethics their way. Had he changed his mind?

"No, Linda – I haven't changed my mind. I'm still on your side of the fence. I'm grateful for anything you give me that will help protect our interests."

Linda toyed with his words in her mind. *Our interests.* She liked that. *I'm grateful for anything you give me...* Yes, that referred to information. It also referred to money. Well, fair was fair. He was on their side, but he was offering to take risks for them. Risk-takers deserved to be paid. She had come prepared. Linda took out a pen, and a piece of plain paper. She wrote a number on the paper, and handed it to Huxley.

"Our 'thanks' to you will show up in this account."

Huxley smiled and took the paper.

* * *

Richard Serrap stood once again at the scene of the accident – and caught himself in the middle of a train of thought. It was no longer the scene of an accident; it was the scene of a homicide. He was here because he needed to think, to brainstorm, to come up with a direction for the investigation that lay before him. He often returned to the scene of a crime to do this. Some people thought better at their desk, or in their shower. Richard thought best at the scene. Besides, he reasoned, the more time spent there, the more familiar he would be with the case. It also increased his chances of seeing or finding something new.

Water-proof boots in place, Richard headed once again for the crest of the little hill, about half a mile from the field. He had his backpacker's chair, which weighed a mere fourteen ounces. His intention was to sit up there to think.

Who? Why? Tomorrow, he would visit Dr. Kline's office and make some initial observations. He planned on making a surprise call upon Dr. Huxley. Richard preferred surprise interviews – he often learned more from what people did not say than from what they did say. And when he made appointments, people tended to be prepared and to talk a lot.

Richard looked through the thin manila folder that he had brought along. This was the folder that hopefully, in the coming weeks, would collect the evidence that pointed to Dr. Kline's murderer. Richard had such a folder for every case he worked. There was the official folder that contained everything. Then there was the little manila folder that contained only copies of the evidence that Richard considered extremely relevant. He would carry it with him always, and it was his way of sorting through details. Everything was important, but Richard's way of solving a case was to locate the main points of the outline, so-to-speak, first. Then, once he had the big picture, he would take the smaller details and see where they fit – if they fit. If they did not fit anywhere, sometimes that meant that the outline had to be revised. But usually, Richard's outline had an uncanny habit of being right-on. What amused Richard, and frustrated some of his colleagues, was Richard's ability to put only the truly "big" things into his little manila folder. Ken had said once that it was as if he reached into a jigsaw puzzle box with his eyes closed, and could consistently pull out only outside-edge pieces. Richard had been amused by this analogy, because he liked putting puzzles together, and he always did the border first.

Richard sat in his little chair, looking down over the field, and then off into the distance where Dr. Kline had spent the last moments of his life, driving towards an unknown death. This was the detective's least favorite part of a case – the point where he had only a handful of edge pieces, and no picture on the box that gave him the slightest clue as to what the puzzle was supposed to look like.

Edge pieces – what were they? So far, the little manila folder contained three items: Christopher Mills' autopsy report, a picture of the tire tracks where the dump truck had been parked and waiting, and the profile of the tracking device that he had found. There was at least one edge piece waiting to be found at 1408 Hamilton Street – Dr. Kline's office building. His widow knew him better than anyone, and her first thoughts had gone to the doctor's professional life. In some cases, this prompted Richard to start looking at secretaries and other female co-workers. Not this time. This murder victim did not fall into that category. The murder was related to the work itself.

This might be tricky. The medical profession had a plethora of laws that they threw up to prevent access to information these days. Some of it was reasonable; some was not. Richard smiled enigmatically. From what little he knew of the deceased, the victim had worked on several occasions to make sure that unreasonable statutes did not prevent law enforcement from doing their job. Dr. Kline had understood that, sometimes, protecting a patient's medical information was not in the patient's best interest. In the past few years, Richard had seen first-hand many broken lives that were the result of minors who had had abortions. Every single one of those cases was an instance of statutory rape. Consensual or not, it was a crime that medical professionals were obligated to report – and didn't. By the time Richard saw the typical case, the abortion clinic and/or the school nurse had stonewalled the parents, who were broken-hearted and often livid over the medical treatment administered to their daughter without their consent. Without fail, Richard's heart always ached the most for the young girl who, by this point, was emotionally shattered. Richard cringed at the consequences due a nation that passed laws which excused and even encouraged sinful choices.

Apparently, Dr. Kline had worked hard from within the medical community to do right by these young girls. Richard had not realized this until he was assigned this case. He wished he had had the opportunity to thank Dr. Kline before he died. Richard would have liked to encourage him to keep trying because, from the detective's seat, the doctor's efforts were vitally needed.

As Richard sat, looking down over the field, his mind ran rabbit trails all around the abortion issue. From what he had learned about Dr. Kline so far, it seemed to be the most obvious politically-charged issue that the doctor was involved in. Richard trusted the widow's comment that suggested that her husband's death was related to a politically-sensitive, work-related decision. Perhaps the decision that the Kline family did not have any particulars about was related to abortion. It was a topic that the doctor had spent decades confronting. He and his wife openly worked with an abortion alternative church ministry in their community. He had lectured frequently on the medical risks to the mother, and had even testified before Congress regarding the issue.

Richard went back over the doctor's personal profile in his mind. Were there any other politically-charged medical issues that were mentioned? No, but certainly others came across his desk. Stem cell research had undoubtedly lain in the doctor's path at one time before his death. What else? Euthanasia? Richard did not think there were any euthanasia laws being seriously considered yet, but... what was that he had read a few months ago about charges of criminal intent being leveled against a nursing home? He needed to research that when he got back to the office.

As Richard went through the laundry list of politically-sensitive medical issues, he began formulating his conversation with Dr. Huxley the next day. Yes, he knew what approach he would take.

*   *   *

"Dr. Huxley, there's a Detective Richard Serrap here to see you."

"Detective? Just a moment."

It was one minute – Richard timed it – before Dr. Huxley's door opened. The man wore the forced smile that Richard was used to. Sometimes people were supposed to smile when they saw him, and they knew it. But Richard knew it, too, and he knew the smile; it never quite reached their eyes. Dr. Huxley graciously extended his hand, and invited him into his office.

Richard sat in the indicated chair as Dr. Huxley spoke the customary greeting, "What can I do for you?"

"I need you to help me learn about the things Dr. Kline was working on just before he died. Specifically, were there any morally-loaded decisions that came across this desk recently? Abortion? Stem cell research? Euthanasia?"

Huxley seemed taken off-guard. "I'm not sure what decisions came off this desk just before I took the chair. I'm still sorting through many files, doing a lot of reading, trying to bring myself up-to-date. I expect it will be at least a few more weeks before I am able to act in the full capacity that this job requires. Right now, the answer to your question is 'no.'"

It was a good answer, Richard thought. He probably did not say anything untrue. But it was, perhaps, a little too rehearsed.

"What's your opinion regarding partial-birth abortion?" Richard needed a question that this guy didn't have a rehearsed answer for.

"It's a gruesome medical procedure." Richard had been watching the doctor as much as listening. He had wanted to see how the doctor gave an impromptu answer. But even this answer seemed a bit prepared.

"That's a truthful answer, of course, Dr. Huxley, but it doesn't tell me where you stand."

"Now that I'm sitting in this Chair, detective, I don't consider it my right to 'stand.' Sitting here means that my decision must be dictated by the evidence that is laying on my desk."

"So you don't think there's such a thing as absolute truth?"

There! Based on the doctor's reaction, Richard knew that he had finally asked a question that Huxley did not have a rehearsed answer to.

"What does that have to do with your stated objective for being here? I had the impression that you were here investigating Dr. Kline and *his* decisions. Was I wrong?"

"No, sir. Dr. Kline is, indeed, the subject of my investigation." Richard was about to say more, but stopped. Huxley had provided the perfect closure for this meeting – much better than the one Richard had prepared. He'd take it. Richard stood and extended his hand. "If you learn of any ethics-based decisions that Dr. Kline was involved in just before his death, I'd appreciate hearing from you." Richard handed him a card, politely nodded, and left.

* * *

Richard sat down in his car and closed his eyes. He replayed his conversation with Huxley almost word-for-word. Then he wrote it down. Then he read it. He had not overtly laid any type of a crime at Dr. Kline's feet; Huxley had been the one to assume that. Huxley had jumped to a conclusion that allowed Richard to avoid the word "homicide" altogether. Richard was glad that his card had only his name and phone number on it. On some occasions, having his job description on it was not wise.

Richard had not decided in advance to steer the conversation towards abortion. He had considered the option, because it was the most glaring issue in Dr. Kline's profile. He had followed through on it because of the need to get the conversation into the realm of the unrehearsed. That need had taught Richard that Huxley had been expecting him. Huxley's reaction to the subject had also taught Richard that Huxley sat on the opposite side of the fence from his predecessor.

Richard took out a sheet of paper, and wrote "Abortion" on it. The single sheet of paper went into the little manila folder.

* * *

Dr. Carl Huxley had been very proud of his prepared answer for the detective. He wished that Linda had not told him anything. The little bit that she had told him had not really *told* him anything; it only served to make him nervous. And that, he knew, detectives were trained to pick up. But this seemed to have gone well. Detective...he looked down at the card... Serrap had apparently been satisfied by his answer about "ethically questionable decisions." But why was the guy investigating Kline? He seemed to be interested in some decision that had been made before Kline died. And yet, by steering the conversation towards abortion, Huxley wondered if the detective knew more than he was letting on.

What had Linda said? "Somebody's already asking questions." But she had not said about what. The obvious implication was that it had something to do with the folder in Huxley's safe – the information which Linda had subtly suggested he "misplace" for a while. Carl had assumed that this translated into "hide it, bury it, for as long as possible." He would have done that anyway – and she knew it. So what *had* she been trying to communicate? Huxley stopped his thoughts, as if in mid-stride. He did not want to know. It was sort of a fun challenge to try to figure out what Linda was saying, but in the interests of self-preservation, he laid the challenge aside. He could not get in trouble for what he did not know. Linda was trying to bait him, to draw him into knowledge that could get him in trouble. Then she would have something to hold over him. He would not be that stupid. No. He would do exactly what he would have done if Linda had never contacted him. He would figure out a way to abort the information that lay in his safe.

What was the abortion-related issue that had been laid on Dr. Kline's desk? Actually, Richard did not even know for sure that it existed. But he needed to start looking for something in particular. If it did not show up, then he would retrace his steps and begin looking for something else. But years of experience, plus the evidence he had so far, combined with a bit of knowledge, made him inclined to believe that the answer to this question not only existed, but had gotten the doctor killed.

Listening carefully to Dr. Carl Huxley's conversations in the coming weeks might be the best way to find an answer to the question. He could get a warrant and gain access to Huxley's files – and would probably do so, eventually – but now was not the time. If he went charging into the office now, he would likely find out what the issue was, but finding the issue was, for Richard, secondary to finding the killer. For now, it was better to just observe. Huxley thought he was investigating Dr. Kline. Let him believe that. Let him act normally, so that he continued to interact with anyone else that may have been involved in the murder.

Who *was* involved? Huxley? Richard was now walking around the field again, thinking. There had to have been at least two people – one to watch for Dr. Kline, and one to drive the dump truck. Or maybe not? Could one person have done it all? Sit in the dump truck, wait with the receiver, when the car came in range, start out, and then smash into Dr. Kline's car, as long as another car is not visible? Richard looked down the road. Such a scenario was probably too risky. The driver of the dump truck could not see far enough down the road from the field to be certain that another car was not a quarter mile back. There were probably at least two people.

Were they hired, or were they the ones who wanted Dr. Kline dead? They had done a very good job. If Richard had not found the receiver – and the odds had been heavily against the find – the death would have been written off as an accident and the case closed. Mills' suicide note would have been believed, and there would not have been any other suspects to look for.

Richard guessed that the killers were probably hired. Novice murderers could be very smart and cover their tracks well; Richard knew this from experience. But this felt more professional. A smart person could have killed Dr. Kline and made it look accidental without having to kill what seemed to be an innocent bystander – Mills. Or was there a connection? Was there any reason to believe that Mills was more to this case than just the person who was fingered as a scapegoat? Richard made a mental note to look for a connection between Mills and the Klines.

Was Huxley involved? He struck Richard as rather arrogant, but not cold and calculating. Huxley was certainly a suspect, but not a looming one. The fact that Huxley apparently disagreed with Dr. Kline's views on abortion led Richard to think that perhaps another party had removed Dr. Kline, knowing that Huxley would end up in his chair. Huxley may be an oblivious pawn.

Who knew what decisions Dr. Kline was making before he made them? Very few. Those people needed to be listened to, too.

Richard's mind went back to the mysterious issue. He wanted to know what it was, but would knowing the specifics really get him any closer to the murders? If it was an abortion issue that got him killed, then Dr. Kline's adamant pro-life position meant that the murders were probably in the other camp. That's who he had to look for:

a pro-abortion advocate who 1. knew what issues were laid at Dr. Kline's feet before they were made public; 2. knew that Huxley would take Dr. Kline's chair, or who could place him in it; and, 3. was willing to kill someone in order to make sure that it was Huxley, and not Kline, who made the final decision.

Richard sighed to himself, as if recognizing a new piece of information: whatever the mysterious issue was, Richard would probably know soon enough. Huxley would make the decision, and then it would hit the streets as public knowledge. Or would he? With Kline's death, maybe he would sit on it for a while, lest attention be drawn to the difference in position between the old chair and the new chair. What had Huxley said? "I'm still sorting through many files, doing a lot of reading, trying to bring myself up-to-date. I expect it will be at least a few more weeks before I am able to act in the full capacity that this job requires." Yea, he was going to sit on it for a while. He may or may not have been involved in the murder, but he was, at the very least, either manipulating or being manipulated.

*   *   *

Linda Frederick liked this meeting. It made her feel important. The man sitting across the table from her did not have many meetings on his calendar. He had enough power that he usually let others do that kind of leg work. But when the war called for hundreds of millions of dollars worth of help, it was his signature that accessed the money.

"As I see it, this meeting is now totally unnecessary. You have thirty seconds to make me believe otherwise."

Linda knew he was not bluffing on the time limit, and knew better than to use much more than half of it. "You were prepared to spend a hundred million three months ago on the plan that never made it to the floor. I need half that much in order to guarantee you $750 million more in tax money. Fair deal."

His dark eyes squinted and glared a little bit. In spite of the expression, Linda was struck, not for the first time, by how handsome he was in a white shirt and a suit that would have bought a third-world resident meals for the rest of his life. "I'll watch how your Committee Meeting goes." Then he got up and walked away.

Linda knew this meant that, if she came through in Committee, she'd get a nice reward. Probably not the whole 50, but plenty to make her happy. She'd have to share it with Huxley, but he would not know how much she received. As long as his bone had a bit of marrow in it, she could keep most of it. Huxley did not have to be paid to change or keep his ideology, and he did not want to take risks – that made his retainer cheap. Linda smiled ever so slightly to herself as she left the plush room. Years ago, she had helped put Huxley in the queue for the position he now occupied, and she was proud of her forethought. The timing had been perfect.

9

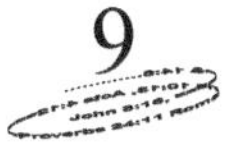

Richard sat and held Susan so very tenderly, as though she were a porcelain doll that might break. She didn't feel well again, and he knew that it brought her some comfort just to be held. She often said that it was the most special thing he could do for her.

Richard had determined in his heart to spend all day Saturday doing nothing else but ministering to his wife. He had made her breakfast. She liked oatmeal with fresh fruit, so he had cut up an apple into tiny chunks and put half of it in the bowl with about half a cup of cereal. It looked like so little food, and yet Susan had worked hard to finish it.

Then she went back to bed and rested while he cleaned up. When he came in the bedroom a little later, she was asleep, so he sat beside the bed and read, watching for when she woke up. He offered to read to her then, which she liked. They did a Bible lesson first, and then she took another brief nap. Then he read her some highlights from the newspaper. Richard marveled at his wife. She would probably only be alive a few more months, and yet she was interested in knowing what grieved her country. Richard knew that during the many hours when Susan's weak body was confined her to bed, she prayed for her nation. That was why she wanted him to read the newspaper to her. He also tried to always find a humorous story to close with. He had found a great one today about a car thief who called the police to remove a boa constrictor from the car he had stolen. Unbeknownst to him, the car's owner had lost his pet in the car earlier that day. When he reported it stolen, he explained that the snake was in there when the car had been taken. To Richard, it was the most beautiful music in the world when Susan laughed. He had always thought this

– from the moment they had met twenty years ago, he thought her laugh was beautiful. But these days, the sound was more precious than ever.

Susan took another rest after their newspaper reading while Richard made her lunch. The salad was a work of art. He tried so hard to make it beautiful to the eye, because he knew Susan's appetite was weak at best. His efforts were not wasted. Susan smiled and truly appreciated the artwork, because it really was easier for her to eat food that looked nice.

Susan insisted on helping to clean up lunch, despite Richard's protests. She got her plate and his to the sink, and then the world began to spin. She grabbed the counter, and would have fallen except that Richard was watching her carefully. He was right there to uphold her and, when he realized that it was necessary, he picked her up and carried her back to bed. His heart constricted, and the tears threatened to spill. She weighed so little. By the time he had gingerly deposited her in their bed, he had his emotions back under control, and managed a tender smile. "You rest." His voice was almost a whisper. "I'll be back in a little while." Susan smiled back, but her eyes were closed.

*   *   *

Carl Huxley was not thrilled with the idea of being in the office on a Saturday. But he had needed to come when he knew he would have the building pretty much to himself. He had read the contents of the manila folder twice now. The first time had been a week ago. Within two days of assuming Dr. Kline's position, he had read all the cases that needed to be decided. When he saw this one, he had taken a deep breath, rubbed his eyes, and then sat for several minutes with his head in his hands. In fact, he had almost declined the position because of it. It was going to be a hot potato that remained hot – which meant that someone was going to get burned. Since it was sitting on this desk, the burn victim was probably going to be the person in this chair. He didn't want that. But if he was not the one sitting here, he was not sure who would be – and this decision was far too important to leave to chance. It was bad enough that Kline had almost been the one to decide. A week ago, Huxley had thought that the timing of Kline's death had been fortuitously lucky.

Today, Huxley had read the entire file again. Then he swiveled his fancy leather chair around to face the window. Fortuitously lucky? Maybe. Or not. For the second time, Huxley seriously considered walking out of this office and letting someone else have it.

*   *   *

Richard got lunch cleaned up, and went back in the bedroom. Susan was sitting up. They had bought one of those beds that sat up when you hit the button, and her side was independent from his side. This was very practical, he knew. But sometimes, like now, he would feel frustrated by the picture it presented: independence from one another. The cancer was probably going to separate them here on earth, and the fact that it had invaded their bedroom and already necessitated this little "separation" seemed like a picture of what was to come. Richard dismissed it as a silly thought, but in truth it was a natural reaction from the heart of a man who really loved his wife.

Richard pushed the button on his side of the bed until it was in the same position as Susan's side. This allowed him to sit next to her.

"Tell me how work is going for you."

"Remember the car accident a few weeks ago, where the medical doctor was killed?"

"Yes." She had been praying for the widow frequently.

"It wasn't an accident. The man turned out to hold a position with a great deal of authority in the world of medicine. It appears that he was about to make a very unpopular decision, and that someone wanted him replaced before that decision was made public." Richard paused, weighing how much information he wanted to share with Susan.

"A decision about what?"

* * *

Huxley swiveled the chair back around and stared at the folder, as if it might bite him. It was so tempting to just walk out the door, and leave the folder right there. Whoever walked in the door after him would understand. They might even follow him out the door, too. But no. If he ran out on this decision, and the person after him decided as Kline would have, Huxley would not want to live with himself. Kline had had his convictions, and Huxley had his. This was, quite literally, the opportunity of a lifetime. To pass this off to someone else would be...like abdicating a kingship – not for the sake of love, but rather out of cowardice.

Then another thought struck the doctor: If Kline's death had not been an accident, and he did walk out the door, what would happen to him? The blood drained a little from his face, and his heart pounded in his ears as he realized the possibility that he had been placed here with a purpose. If he thwarted that purpose, would it cost him his life?

Huxley shook his head slightly. He *wanted* to make the decision that those powerful people agreed with. It certainly was not worth risking his life when he actually agreed with what they wanted him to do. Nevertheless, the feeling that settled in the pit of his stomach was not pleasant.

* * *

Richard was silent for just a moment. He could say that he didn't know the subject of Dr. Kline's mystery decision. That would be the truth; he really wasn't sure. But Susan knew him too well. She would be able to tell that he was avoiding her question. Besides, Susan had told him on a number of occasions that she was able to talk about her past. Nevertheless, Richard avoided even allusions to the painful subject. She was dealing with enough pain in the present. He didn't want to add to it. "I'm not sure, but I think the doctor's decision was abortion-related."

The sigh from Susan's heart was as audible to Richard as the one from her lungs. He understood, and squeezed her hand just a little.

"How many, Rich? How many have been deceived like I was? How many more children will die?"

"Susan...don't."

"No, Rich. I know – you didn't want to tell me because you didn't want me to be reminded. And that means a lot to me. You appreciate that my heart still regrets my choice. But by the grace of God, the pain is gone. I can hardly believe that I'm able to say that. As you know, I had lost hope that I could escape the guilt, and the pain almost killed me. The scar is still there, but when I obtained forgiveness, true healing began. At that point, what the enemy intended for evil, God was able to begin using for good. It's been a blessing, Rich," she took his hand, squeezed it, and looked into his eyes in order to emphasize her words, "a *blessing*, to be able to spend all those years helping those girls I counseled. And I praise the Lord for every minute that He gives me the strength to pray against that horrible deception."

Richard squeezed her hand. "I love you." He could not imagine living without this woman.

* * *

Huxley swiveled his chair around again and looked out the window. He would stay. He would make the decision he wanted to make – on his own. He did not want it to look like he was paid to make this decision. There was no use risking the trouble that such appearances might cause in the long run.

* * *

"I love you, too, Richard Serrap." She may not feel perfectly healthy, but she still loved to kiss and be held by this handsome police detective that she was blessed to call husband.

* * *

"Linda, this is Carl. I've made a decision. You know where I stand, and I'm not going to change my mind. But I don't want to be paid for a job that I would do anyway."

At first, Linda was livid that he had left those words on a message. Then she smiled. Now they were deleted, gone forever. It wasn't such a big deal. And if Carl didn't want his share of the money, she wouldn't argue.

10

"Hey, Chief! Wait 'til you hear the message we recorded this weekend!"

"What do you have, Charlie?"

Charlie hit a button... "Linda, this is Carl. I've made a decision. You know where I stand, and I'm not going to change my mind. But I don't want to be paid for a job that I would do anyway."

Richard was, indeed, pleased. "Linda who?"

"Linda Frederick. Recognize the name?"

The expression on Richard's face was enigmatic. "Our very own Representative."

"I'm impressed. Most people don't know their Representative's name."

"My wife and I have written her more letters than I can count. And we pray for her often."

*   *   *

Once again, Richard sat in his backpacker's chair at the scene of the crime. This time, though, he was not up on the hill. He was down in the field, near to where he had found the receiver. He looked through the manila folder. It now had four edge pieces: Christopher Mills' autopsy report, a picture of the tire tracks where the dump truck had been parked and waiting, the profile of the tracking device, and the word "abortion." Richard took out another piece of paper and wrote "Names" at the top. This was typical; often, the thin manila folder contained such a sheet, and people were added to it as the case progressed. Near the top he wrote Linda Frederick's name. Down at the bottom, he wrote Carl Huxley's – then crossed it off. Huxley was a pawn. Richard was nearly certain of this, and so he did not belong on this sheet. He drew an oval above Frederick's name, and put a question mark in it. There was at least one more key person whose name he needed to learn.

Richard put a dollar sign in the oval. That was the key. The missing person was the money behind the crime. Richard was nearly certain that the job had been hired out, and it was done well enough that it had been an expensive job. Criminals who often left no physical evidence frequently left a money trail. Richard knew that such trails often contained much information. How much money had been spent was, in itself, very important. Where it was saved, where it was spent, and who ended up with it often indicated where it had come from.

Where was the money? By now, it was likely in the hands of whoever had executed the plan. How much? That was the question Richard was the most interested in right now. Finding one or two names was valuable, but it was often the dollar amount that helped Richard determine if they had found the person at the top. He wanted to know how high he ought to be looking.

Richard's cell phone rang. "Serrap."

"Richard, it's Jim. I've gone over Mills' body with a fine-tooth comb. As I said before, I think I have enough evidence to convince a jury that someone else's hand was on Mills' hand when the trigger was pulled. But they must have been wearing gloves, because I can't get a trace of them. I did find evidence of alcohol. Beyond a reasonable doubt, he was drunk when he died. However, for the sake of playing devil's advocate, if someone claims that they were trying to *prevent* a drunk, depressed kid from pulling the trigger, I can't give you any evidence to the contrary."

"Good work, Jim."

"I'm sorry I can't give you more."

"That's OK. More than anything, I want the truth. Talk to you later."

*   *   *

Richard opened the creaky door to a run-down cabin in the middle of the woods. It was actually sort of nice, for a cabin. The floor was wood; there were a couple of good windows, a rough table, a couple chairs, and a cot. A big fireplace could be used for cooking and would easily heat the small 12' x 12' building, in spite of the draftiness. There was even a small counter and sink, with an old hand-operated water pump. Richard was impressed to discover that it tapped into cold, great-tasting water.

This was where Mills had been found, several days after Dr. Kline's death. Richard pulled out his map with the red "X"s, and laid it on the old table. He added a new one where this cabin was located. He squinted, thinking about the route that a young man on foot would have taken to get here from the location of the doctor's death. It didn't work. First of all, it was too far. As soon as the then-accident had been discovered, the search for a suspect on foot had begun. It was almost impossible that he had come directly here; they would have found him on the way. Second, there was the alcohol that he supposedly got himself drunk on. Perhaps that was already here, but Richard thought this was doubtful. Third, there was the suicide note. Richard looked around. There were no books, no paper or pens. Those things almost certainly had been brought here after Kline's death. Fourth, there was the gun. That, too, was almost certainly brought here after Kline's death.

Could Mills have had the gun with him when he stole the truck? Richard scowled as he considered this possibility. He really didn't think so. The picture that someone had tried hard to paint was that a young, reckless kid had stolen a dump truck for fun. Then he lost control of it while going on a joy ride and accidentally killed someone. He fled the scene of the accident, probably aware of what the doctor's dead body had looked like. He hid out for a few days, scared. Finally, consumed with guilt and perhaps the horror of the tragedy, he had gotten drunk and killed himself. The idea that such a kid would carry a gun when he initially stole the truck didn't track. The gun must have been obtained after the crash, along with some of the other items.

Still, a bit of twisted logic could easily fit the gun and other items into the initial story. Distraught and afraid, the young man goes back to his house. He knows he can't stay there, and remembers the cabin. Stop. Richard had been walking around the little cabin, and suddenly he froze. Once Mills made it back to his house, why wasn't he safe? At that point, no one had even known that he was the one who had stolen the truck. As far as Mills had known, he had totally gotten away with stealing a dump truck out from under the nose of the county...

Richard walked over to the table, sat down, and closed his eyes. In his mind, he slowly and deliberately "read" the account of the truck theft. Apparently, Mills had brazenly walked into the yard and right over to the dump truck. The keys happened to have been in it, he started it, and drove out of the yard with it. Only one witness had seen him and, at the time, was a bit suspicious. But the witness was new and did not know everyone who worked there yet. The young man had seemed so confident that the witness just assumed that he was wrong to be suspicious, and continued his job. It was a couple of hours later that someone in authority noticed that the dump truck was unaccounted for. It didn't take long to find the witness, who got dressed down for not having said anything earlier. But he swore that the young man acted just as though he were assigned to take the truck. Besides, he thought that he had seen him around the yard before. How was he to know that the kid was stealing the truck?

Richard considered this. Yes, it was feasible. He remembered one case where he had caught a bunch of boys who had become very good thieves by virtue of acting confident, just as Mills had. In fact, they had successfully walked into a furniture store and managed to walk out with a couch without getting caught at the time. Richard knew that security at the county yard wasn't spectacular, so yes, it was possible that Mills' cockiness had allowed him to simply drive away with the truck.

Richard continued through the report in his mind... The truck was reported as stolen, but there was no named suspect. The witness had said the young man's face "was familiar," but could not identify him. Later that day, Kline was killed. The witness was immediately paired with a portrait artist, since his identity was now more crucial. But that yielded no immediate results. It wasn't until the next day that Mills' parents reported him missing. The witness was then shown a picture album of thirty young men, and fingered Mills with a high level of certainty.

It was a good story. It made sense.

But was something wrong? Before the accident, Mills probably thought that he had gotten away with taking the truck. From the way the witness spoke, Mills probably did not even see him, and no one else was around. After the crash, Mills ran from the scene. Once he was clear of the area, what made him think that anyone could connect him to the dump truck?

Richard shook his head. Here, now, looking back, it was easy enough for Richard to see that there was nothing connecting Mills to the truck at that point. But the kid was probably scared senseless, and his own guilt would have been enough to motivate him to hide, even if he had been aware of the lack of evidence. At that point, he probably felt sure that "they would find him." Consequently, he couldn't stay at home. He thought of the cabin. It was the kind of place that many of the teenagers in the area probably knew about – or did they? Again, Richard stopped. He looked around. It was nice, for a cabin. Not vandalized. On second thought, most of the local teenagers probably did not know about this place. So how did Mills?

* * *

Dr. Huxley enjoyed parking his car Wednesday morning. He now had a parking space that was his alone. Such a little thing – in a way. But he liked it. Having made the big decision the previous Saturday, the rest of the weekend had gone quite well. When he had come to work Monday morning, he was able to pretty much forget about the folder in his office safe. It was going to just sit there for a while. On Tuesday, he had gone about his work and ignored the invisible thing.

Walking into the office and sitting down at his desk, Dr. Huxley began the day's work. The phone rang, and he let his secretary pick it up. "Dr. Huxley, Dr. Jacob Davidson is on line 1."

"Tell him I'll have to call him back."

Great! The primary author of that accursed file! Huxley knew what he was going to say... "Did you read it yet? What are you going to do? There are millions of lives at stake here; we must do something now! You can't sweep this under the rug." Yes, Huxley knew what Davidson would say. He sighed, and wiped his hand over his face. Then he leaned back and rubbed his forehead with his hand. He should have known better than to think he could leave this thing buried in his safe for long. It was almost as if the thing were alive and growing; it would not tolerate being totally hidden for long. And the longer he waited, the harder it was going to be to abort it.

Maybe he had made a mistake by thinking that he could ignore it for a while. That route would only give it time to grow, and make it harder to kill. Furthermore, the longer he tried to hide it, the more obvious it became that he was trying to destroy it. The solution, he now realized, was exactly the opposite of his original intent: throw it out into the public eye before it was mature. As it existed now, the data in the file was incomplete; it was not yet ready to stand up to the scrutiny of a court of law. Given time, it would – but not yet. So, force it into the courts now. Its own immaturity would be its demise.

Some people would scream and holler, saying that more research should be done before the issue went to court; that the data needed to be mature before it was forced to fight for its life. But Huxley would turn that logic right around against them: those were the very people who would want this information publicized to the hilt. They couldn't have it both ways. If they wanted it publicized, then the courts had to determine its legitimacy; if they were not ready to allow the courts to determine its legitimacy, then it was not right to publicize it.

Of course, Huxley knew how they would respond to that. They would claim that the public had the right to know that the connection existed, and it did not matter that the complete explanation for the connection was yet unexplained. It existed, and it was only a matter of time before all the details could be explained. Since it existed – even though it could not be fully explained or understood – medical and legal decisions must reflect its existence.

Huxley had an uncomfortable feeling in his stomach as he perceived the logic of the other side's argument. He purposely suppressed it. *He* was in this chair. *He* was in charge. He had already made up his mind about what needed to be done with this file, and he wasn't going to change his mind. He did not want to change his mind and, considering what had happened to his predecessor, it would not be in his own best interests to do so.

* * *

Richard was concentrating so deeply on the events surrounding Mills' death that he did not notice the little mouse that ran in through the front door, and scurried over to the corner where the cot was. He was sitting at the table, eyes closed, still going over the events that led to the discovery of Mills' body.

Once the witness at the county yard had identified Mills, an intense search for the young man had ensued. His parents cooperated as much as was possible, but they did not know where their son had disappeared to. Richard was inclined to believe that his friends from school were telling the truth when they claimed ignorance of his whereabouts.

At this point, Richard began to review the character sketch. Christopher Mills had been 17, a high school senior. He was not always very responsible, and had been known to pull some pretty foolish pranks. For a while, he had experimented with drugs, but apparently had stayed away from that crowd over the past year. In fact, with the addition of a new girlfriend into his life a couple of months ago, it appeared to all the world that he was on track to a level of maturity commensurate with his age. When his theft of the truck had been presented to those who knew him, the response was mixed. Some were not surprised; Christopher still did some stupid things, especially when alcohol was involved. But others hesitantly told Richard that the news "just didn't seem right." A couple of years ago, they would have believed it, no problem. But the accusation did not track with the young man's more recent behavior. The girlfriend was among those who did not believe that Mills would do such a thing. In fact, they had apparently made plans for the following weekend. Christopher had promised her some sort of special surprise. But, from Richard's perspective, she had not known him long, and did not seem to know him very well yet.

Richard had never found evidence to indicate that Mills had driven any vehicle towards the cabin, nor was any witness found who had given the young man a ride. It appeared that he had gone the whole way on foot. This was necessary for the last mile, which was a mere foot trail. But there were several miles in between Mills' house and the cabin that could have been driven. Richard shook his head. The way they had been searching, there must have been a car, or a four-wheeler, or something.

The cabin had been located from the air four days after Dr. Kline's death. When it was checked, Mills' body had been found inside. A single gunshot had immediately killed him, and the suicide note expressed deep sorrow and guilt over the accidental death of "the guy in the car." The coroner estimated his death to be two days after the accident. Two liquor bottles, both empty, were in the cabin. There was also some food. The food appeared to have been there for some time, and Christopher had eaten a couple of meals' worth of it.

The suicide note was bothering Richard. Where had the paper come from? This really smelled like a homicide, but whoever had planned it did a very good job. They would have created an explanation for where the paper and pen came from. Richard looked around, slowly and deliberately.

A movement caught Richard's eye, and commanded his immediate attention. It was the mouse. Richard got up, herded him towards the door, and stepped outside. He looked around. Was the paper their mistake?

* * *

"Dr. Davidson, this is Dr. Huxley. I'm sorry I couldn't take your call earlier. What can I do for you?"

"Have you read our report?"

"Yes."

"What are you going to do with it?"

"I think it is very important research, and the public needs to be made aware of it as soon as possible. I've already prepared a preliminary public statement, encouraging women to look into whether they have been victimized by lack of knowledge." He didn't mention that this statement had only been typed in the last 30 minutes.

"Dr. Huxley, I'm pleased that you recognize the need to publicize our findings, but I think it's too soon to use such strong language. We do not have all the answers yet. The connection is certainly there, but we need more research before we will understand it thoroughly. If we go out with guns a-blazing, it will invite a landslide of immediate lawsuits. It's too soon for that; I'm concerned that we'll run out of bullets, so to speak, before the trials are over. Consequently, the courts will rule against us. Then, many who are truly liable will literally get away with murder: they will have been tried and exonerated, and so once the evidence is mature, it will be unconstitutional to re-try the cases."

"Dr. Davidson, your research clearly indicates that the women are in jeopardy. From this chair, it is my responsibility to make sure that they know it. I can not, in good conscience, hide such a thing!"

"I agree! But it is wrong to overstate our case. The public must be told of the dangerous connection immediately. But if we take it to court too soon and lose, then the public will come away with the mistaken impression that the connection is not really there!"

"Dr. Davidson, you are asking me to ask the public to believe something that does not yet have enough legitimacy to stand up to the scrutiny of a court of law?"

"No. I am asking you to tell the public what we found. Nothing less, *and nothing more*." Dr. Jacob Davidson was now angry. When he learned that Huxley had replaced Dr. Kline, he was sure that he was in for a war; Huxley's apparent decision to ignore the report for two weeks had seemed to confirm this. Then, at the beginning of this conversation, Jacob had been taken aback – perhaps he had misjudged where Huxley's loyalties lay. But now he saw the truth. He was, as he had initially suspected, in for a war. Huxley was not on his side. Dr. Kline's replacement was being very crafty; his intention was to bury the report as deeply as he could beneath misrepresentation. Eventually, the facts would arise – they could not be killed forever. But this strategy *could* succeed in suppressing them in the mind of the culture for quite a while.

*   *   *

Richard, standing in the doorway of the little cabin, looked around the inside of it once more. His eyes fell on the cot. He lifted up the mattress. There was a partial pad of paper. Richard nodded. It was going to match the paper that the note had been written on.

11

The conversation between doctors Davidson and Huxley had ended. Jacob Davidson sat at his desk, eyes closed, with his head resting on his hands. If only Edward Kline were still alive! It was going to be a life-and-death struggle to safely birth the research file that sat in Huxley's office. Huxley was determined to kill it, and with the speed at which he was now moving, he might succeed, at least, temporarily.

Dr. Davidson's thoughts returned to Dr. Kline. He had been a very well-respected doctor, and for good reasons. Davidson had rejoiced last year when he made a break-through, and realized that Dr. Kline would be the one to review it. It was a sensitive, politically-sensitive finding. This meant that it was fragile, and would need special care at first. Dr. Kline was so gifted at caring for fragile infants. Davidson had known that there would be no one better able to ensure the life of his research.

*   *   *

Richard sat back down at the table in the little cabin, and pulled out his manila folder. Something else about Mills' death needed to be added to it. But what?

What was the level of certainty for the handwriting on the suicide note? Very good. Mills' parents had seen the note, and had confirmed that it was their son's writing. Richard had also had it examined by an expert, who confirmed that it was very likely written by Christopher's hand.

If the note was not genuine, was it faked by the hand of another, or coerced from Mill's hand? Richard doubted that it was coerced. Even if Mills had been drunk, he could not imagine such a long note to be the product of coercion.

Was it obviously the wording of a suicide note? The thought intrigued Richard for a moment. Was it the kind of wording that could have been coerced from a drunken person without them realizing that it sounded like a suicide note? The body of the note was made up of a confession and an apology. Richard was doubtful. The wording sounded too much like a suicide note. And besides, if Richard's guess was correct, Mills was not even in the dump truck when Dr. Kline was killed. This would mean that – if the note were coerced – Mills was forced to confess to something that he was not responsible for in what sounded very much like a suicide note. Richard doubted that this was likely.

If the Mills death was a homicide, the note was probably forged. And if it was forged, it was a very good forgery. This meant that it was a very expensive forgery. Very, very expensive.

Richard pulled out the Name Sheet from his manila folder. Underneath the circle with the dollar sign, he drew several dark, bold underlines. Richard was beginning to get a notion of how much money had been involved in the execution of these crimes: a lot.

No. Wait. There was something amiss in the mental scenario Richard had just followed. It appeared that Mills *had* been in the dump truck when it crashed into the car. The young man's body had sustained bruises consistent with the timing of the motor vehicle crash. Richard paused. Was Mills in the truck, and someone else driving? If he had been conscious, he might have tried to prevent the crash. If he had been unconscious, it would have meant that the driver had to carry him away from the scene. Or, were the bruises inflicted later just to make a well-fabricated story consistent? On top of all of this,

Richard still was not 100% convinced of the boy's innocence. That the suicide was genuine could not be totally discounted yet.

*   *   *

When Richard returned home, his mother-in-law greeted him at the door. Richard was pleased to see her. He had never liked mother-in-law jokes, because he had loved Sue's Mom from the moment he met her. Mrs. Sheldon had become a second mother to him, and he loved and respected her a great deal. He was so thankful, for both his and Susan's sakes, that they were able to live within miles of one another.

"Hi, Mom!" Richard hugged Mrs. Sheldon, set his briefcase down, and began taking off his shoes by the door, as he had always done. "Is Susan resting?"

"Yes." The tone in the older woman's voice made Richard stop untying his shoe in mid-action and look up. "It's OK, Richard. She had a very hard day, though. She called me this morning about ten and asked if I could come over. She had so hoped that, now that the treatments are over, she would have a little strength and energy. But she feels like a wrung-out dishrag. The weakness wasn't why she called me, though. She's discouraged because she doesn't feel like her old self."

Richard removed his shoe and struggled against the temptation to throw it. He was tired, but felt that he had no right to be bothered by such a small complaint, considering what his wife was going through. And yet, he was wise enough to know better than to deny the fact. Because of Susan's illness, they had both learned a lot about perspectives and priorities the last few years. When it became apparent that Susan's cancer would probably kill her, Richard had tried, for a few months, to act as though none of his own problems bothered him. What right did he have to be frustrated over a sore foot – or any other such minor problem – when his wife was dying? Susan had seen what he was doing, though. She understood. But she firmly and lovingly made it clear to him that no one believed that his problems had gone away, and if he continued to try to act as though they had, he was lying. With a smile, and that laugh that he loved, she had told him, "Besides, you're not doing a very good job of it anyway – acting as though all your problems have suddenly gone away."

And so, Richard now sat by his door, one shoe off, one shoe on. He was very tired; he had not slept at all the night before. This had made his day at work hard, even though, from the outside looking in, he had made some successful headway in the Kline/Mills case. On top of being tired, his wife was dying. On top of that, she had had a hard day, and he wasn't here. Not that she, or anyone else, expected him to be. And he was so very thankful that her mom had been able to come. But none of that changed the fact that *he* wanted to be here for her. Mom had said that Sue was discouraged, and she did not get that way often. That meant that, compared to the last few weeks, she had had a very, very hard day.

Richard overcame the temptation to throw the shoe, and just dropped it. He leaned over, jaw clenched, and put his head in his hands. Mrs. Sheldon was a wise and loving woman, and could see the emotions rolling through her son-in-law just as well as if they were physically visible. She put her hand gently on his shoulder, and did not say a word out loud to him. Instead, her heart fervently petitioned the Lord for strength for this man who was just as much her son as if he had been physically born to her.

The moment of grief passed, and Mrs. Sheldon removed her hand and walked toward the kitchen. Turning at the door, she addressed Richard in a kind, gentle tone.

"Susan's sound asleep. When you want, why don't you come on into the kitchen?"

Richard smiled. How could this woman be so perfectly aware of how best to help him? He smiled again, thankful. He knew the answer to that question. He walked into the kitchen, where Mom was putting a lid on a pot. She turned.

"Your dinner's all set, so you have time to swim or relax before eating. Susan asked you to wake her when you're ready to eat, so that she can join you." Mom put the final touches on her dinner preparations, and turned to go. Richard knew that, now that he was home, she would go ahead back to her house.

"Thanks, Mom. Your help means so much..." Richard stopped speaking, because the gratitude in his heart threatened to spill over and make his voice break.

"You're welcome. We love you guys." And then, with a smile, she left.

*   *   *

The following day, Richard made an appointment with the supervisor of the county yard from which the truck had been stolen. He did not learn anything interesting in their short meeting, but he was more interested in walking around the yard. He walked the route that Mills supposedly walked, at the same time of day. Yes, the story seemed to fit with reality. This time, there was no one in that section of the yard, not even the witness.

There was something bugging Richard. Where had the key to the truck come from? According to policy, a key never should have been left in it. The key was supposed to have gone back on the Board, where all the equipment keys were kept, and locked each night. The Board was open all day, but it was over in the Supervisor's Building, and supposedly Mills had not gone anywhere near there. The last person to operate the truck swore that he had put the key back on the Board. Both Richard and the yard supervisor believed the man.

Richard went back in his mind and considered the key that had come out of the ignition of the wrecked truck. It wasn't a brand-new copy; it was a well-worn key. By all appearances, it was the one from the Board – which was now missing. How had it gotten to the truck? And there was another problem: How did Mills get to the yard? It was too far to walk from his house or school, and there was no car left nearby.

* * *

The hospital room was a flurry of nurses, respiratory therapists, and a couple of doctors. It took the team 20 minutes, but eventually the patient was stable.

"This child's going to die."

"No. She is not." The Dr. Gary Jones' voice was not at all loud, but it was firmly filled with conviction.

"She's *supposed* to be dead."

"Only because of the choice of a few people."

"Including her mother. It was a legal abortion."

The doctor looked at the tiny patient, who did not weigh quite two pounds. "I know." His heart ached with a fierce protectiveness. She was perfect. So tiny. So helpless. In his eyes, beautiful. Most other people would be distracted by the respirator, the monitor wires, and the IV lines. But he looked at his patient as though all these things were invisible. He saw a miniature baby, perfectly formed. Her whole hand was only slightly bigger than the end of his thumb. Right now, she was fighting for her life, but not because anything was "wrong" with her. Her body was simply too immature to live outside her mother's womb without special assistance.

The monitor beeped, and the doctor glanced at the screen. Her heart rate had dropped a little, but recovered on its own. "Good girl." He smiled. It was a normal occurrence in the life of a premature baby, and he was pleased that no intervention had been needed.

"What's the story?"

"Mom began the abortion procedure yesterday, and was told to return to the clinic today. But instead of going to the clinic this morning, she went to an ER. She told the staff why the labor was premature. As soon as the child was born, she was covered under the Federal Born Alive Act."

Underneath a barely audible sigh, the physician wondered how many such babies were born in abortion clinics, and left to die. How ironic: the law of the land recognized their right to life only after someone had unsuccessfully tried to murder them.

Several hours later, the doctor returned to the bedside of the tiny infant. Everything was quiet, except for the sound of the respirator. He noticed that someone had made a little name tag for the bed that said, "Kayla." He looked inquisitively at the nurse. "Did the mother choose a name?"

"No. We were informed that she walked out of the ER, having given a false name and false address. When she was admitted, she said she did not have insurance and was going to pay cash. She's gone. The name is unofficial; it's just how we are referring to her for now."

The physician turned back to his tiny patient. His voice was soft and gentle, but firm. "Hang in there, Kayla. You can make it." She moved her tiny arm, and repositioned her hand so that it was closer to her mouth. He smiled. "Yes. You can make it."

* * *

Linda Frederick was not having a good day. One of her philosophical opponents in Committee had presented her with a press release by Dr. Carl Huxley. She had then been informed that the Committee was going to entertain a motion to cut off federal abortion funding within the Welfare Program. She had begun dialing Huxley as soon as her office door was closed.

"Carl! What in the world do you think you're doing!?!?"

Carl hung up. He regretted even answering in the first place. Linda knew better than to call his office phone, and he had never quite figured out how she had learned his cell phone number. It rang again. He turned off the ringer. He was going to handle this without any influence from that woman. He was going to give her what she wanted, and then "they" would not have any reason to be displeased with him. But he was going to do it legally.

She had left a message. OK, he'd listen. "Carl, I'm sorry to have jumped on you as soon as you picked up. Everybody's frazzled over here. The other camp is elated and running for the presses, and our camp is trying to figure out what you've done. What's the deal? I thought you were on our side? Did you change your mind? Do you want to be paid now?"

Carl laughed out loud to himself, and shook his head. She thought she was so smart and powerful, but she had fallen for his well-thought-out deception. Good. If he had tricked someone on his own side of the fence, hopefully the other camp would step in the trap, too. He liked the idea of having put one over on Linda. She tended to treat him like an underling. Besides caressing his pride, Carl considered that having Linda believe the deception was a great tool for propagating it.

Then he stopped, mid-thought. Linda needed to be corrected. Her powerful friends needed to be reassured that he was still on their side, lest he end up like Edward Kline. Great! Now he needed to call her, and he had told himself he would never do that again.

*   *   *

"It's time for the money to stop. You wouldn't listen to me when the evidence clearly showed that we were paying for murders, plain and simple. You responded, 'The woman has rights!' I ask you, do you still hold to that statement? Because if you do, the very women you claim to be so protective of are the ones who will be killed if our abortion funding continues."

There was silence for a moment, and then Linda stepped into it. "Since when do we alter hundreds of millions of dollars worth of funding on the basis of one unconfirmed medical report? The courts need to deal with the legitimacy of the supposed finding."

"Since when are judges medical experts? Carl Huxley's job exists for the express purpose of providing high-level medical expertise to lawmakers. And this," the speaker dramatically dropped the report in front of Frederick, "is his report. It is our responsibility to act upon it, and there is no legal justification for the courts to touch it!"

Frederick was cornered, and she knew it. She resented Carl Huxley. She was now living the nightmarish scenario that Kline's death was supposed to have prevented. She had needed to talk to him before this meeting, but he had not returned her call. "We'll see how your Committee Meeting goes." Those were the words to which her $50 million were tied, and right now, the Committee Meeting was slipping through her fingers.

"I'm not willing to vote on it today." It was a bluff, sort of, but it was the best she could do without even a pair of fours in her hand. She looked around the table. The two sides were pretty well matched, and if the moderate was gutsy enough to follow her lead, the vote would be stalled. He glanced her way. She knew what he was thinking. He wanted to vote with her, and he would do so if he thought she had an ace up her sleeve. But if he thought she was going to loose the battle, then he'd vote against her.

Linda's opponent spoke. "I move to vote on the continuance of federal abortion funding over the next six months."

"I second that motion."

"All in favor..."

The moderate did not raise his hand. For the time being, federally-paid-for abortions would continue. Linda smiled, pretending she had been totally optimistic. In fact, the moderate apparently had more confidence in her position right now than she did. What was Huxley doing?!?

*   *   *

"How's Kayla?"

"All things considered, she is doing very well. Without any major surprises, there's a good chance that she will not only survive, but lead a perfectly normal life."

*   *   *

Richard was back at the county yard again. Today he was going to be Mills. He left his car outside the yard, and walked in. Question #1. He still did not know how the young man had gotten there. He walked to where the truck had been parked. Mentally, he got in, started it, and began driving it out of the yard. The key was not supposed to have been in it. If it had been there, its presence was likely a part of the elaborate homicide. What if the key had not been there? This was supposed to have been a prank – the kind of prank that fell apart without a readily available key.

Mills' actions and described attitude pointed to the fact that he either possessed the key or expected it to be in the truck. Question #2. How did Mills come into possession of a key that he apparently expected to have?

Richard began walking back towards his car. Mills was driving the truck now. Was this reasonable? Yes, the boy had had some experience last year driving heavy equipment. Richard began driving to where the truck had been parked prior to the accident. Once there, he got out and looked around. He looked at his watch. Too much time. The truck was probably not brought directly here. There were more than five empty hours before the accident, and if the truck had been waiting here all that time, someone might have noticed it. The tracking device had made Kline's schedule predictable, and there was no reason to be waiting here for more than 30 minutes.

So where had the truck gone for those two and half hours? Towards the cabin! Richard got back in his car, and returned to the yard. From there, he took the most logical route towards the cabin. He stopped at the point in the road which was closest to the cabin, where he had parked before. From here, the cabin was a mile away, on foot. Then he kept driving. A mile farther up the road he stopped again. Muddy tracks from large equipment were noticeable on the blacktop. Just a little ways off the road, someone had been dumping loads of what appeared to be clean fill.

Richard got out and began walking around. Perhaps Mills had brought the truck here...suddenly Richard had not only an edge piece, but a corner. Suppose Mills had been asked – or even hired under false pretenses – to bring the truck here? Someone had driven him to the yard, and given him the key. Then he brought the truck here, just as he had been instructed. To the young man, it may have seemed perfectly legitimate. He arrives here, and is met by at least two people. One takes him to the cabin; the other takes the truck, perhaps waiting here awhile first.

12

Carl Huxley began dialing Linda's number. Then he stopped. It felt cowardly to call her, because the only reason he was doing so was to insure the protection of his own skin. He wanted to make sure that her friends knew for certain that he was still on their side.

But did he really have to call? Linda might be a little slow figuring out what he was up to, but those above her were probably much more shrewd. They probably saw his plan the moment the press release came out – and admired him for it. So why did he have to call her again, when he really did not want to anyway? Calling her just left a trail, and made him look like her pawn.

He hung up the phone.

* * *

"Why does she claim that she wants to see me this time?"

"She wants to be instructed on what you want her to do about the new problem."

"What new problem?"

"Huxley."

"Like I said, 'What new problem?'"

"She claims that he's turned against her."

"No, he hasn't. Tell her to read the fine print." And then, half-spoken under his breath, words meant only for the ears of the messenger, "He's smarter than she is."

* * *

"What happened!?!?" The 2 a.m. phone call had informed Dr. Jones that Kayla was on her way to surgery.

He did not need to be present, but he had asked to be informed of any major changes.

"Her intestine perforated." It was a common problem with preemies and, although it posed a serious threat, the physician was relieved. Of all the things that could go wrong and send her to surgery at this point, this was one of the less serious ones.

"Thanks for calling. Do you need my help?"

"Not this time, but I appreciate the offer. Good night. Hope you can get back to sleep."

* * *

This was one of those times when Dr. Jacob Davidson was glad that he was still single. He had called his research team together, and gave them a short briefing. Huxley was out to kill the impact of their research by forcing it into the courts prematurely. They had to complete the next phase of the project – which had been originally slated to last more than a year – within months. This thing was going to hit the courts within months, if not weeks. The slowness of the judicial process would buy them some crucial time, but it was still going to be a life-and-death race. There was one piece of critical evidence that had to be found before Phase Two would be complete, and Davidson was prepared to live, breath, eat and sleep on the problem until he solved it.

* * *

Linda Frederick was offended by the terse words and condescending tone the messenger had used. "He says it's not a problem, and that you should read the fine print." And that was all he had said before he hung up.

She was still wondering what that message had meant an hour later when she picked up a copy of the opposition's response to Huxley's release. What was this?

The title implied overt criticism. She picked up her coffee mug and began reading. The article was accusing Huxley of overstating the evidence and laying a trap of premature malpractice suits that were actually designed to protect the abortion clinics! Now she really resented him! All he had to do was explain himself on the phone the other day, and then she would have enjoyed the Committee Meeting and, more importantly, would not have made a fool of herself earlier today.

*   *   *

Richard's first glance at the front page of the newspaper told him why Edward Kline had been killed. In large, 72-point font, the headline below the crease read: "New Research Might Link Abortion and Breast Cancer." So this was it. This was the issue that Kline's enemies were willing to kill him over – which meant that "might" was not nearly a strong enough word to describe the weight of the evidence. Richard put a copy of the article in his manila folder. The edge pieces were just about all there now.

Richard laid out his edge pieces: Christopher Mills' autopsy report, a picture of the tire tracks where the dump truck had been parked and waiting, the profile of the tracking device, and the word "abortion." Stapled to the last page with that one word was a copy of the newspaper article. Then there was the "Names" page. Near the top was an oval with a dollar sign in it. Under the oval, connected with a line, was Linda Frederick's name. At the bottom was Huxley's crossed-off name.

Richard picked up his pen. He needed to add to his "Names" page. Next to Linda's name he wrote "John Minoth." John Minoth, he was convinced, had hired young Mills to take the dump truck to the lot with clean fill. But Richard was not sure who had hired Minoth. Minoth's name was connected with a dotted line and a question mark to Linda's name, and then another dotted line with a question mark connected Minoth to the money bubble.

Minoth was the witness – the new guy at the County yard. A few weeks before Kline's death, he had applied for a job. An opening was available, and the man was not only willing to accept a mediocre wage, but he was available at all hours, including weekends and holidays. Finding someone who didn't complain about the idea of salting the road on Christmas day was a challenge. Minoth was single, and didn't seem to care.

Richard was pretty certain the man had been a plant. He didn't have to work at the yard too long before he knew the routine. Lots of the guys went into town for lunch. On the day of the theft, no one could remember if Minoth had done so. He had kept pretty much to himself ever since he was hired.

Minoth could have approached Mills at some point and offered to hire him for a job. The young man might have been suspicious about the idea of being asked to drive county equipment, but not necessarily. Minoth was an employee there.

Once Mills agreed to the job, Minoth could have picked him up when it would have appeared that he was going out for lunch. When they returned, Minoth could have parked outside the yard, as some of the employees often did. He could have already taken the key off the Board and either given it to Mills or else left it in the truck. And then everything that he had said about what he saw that day was, technically, true. He had seen the boy walk into the yard, get in the truck with such confidence that he looked like he belonged, start it, and drive away. Even the fact that the boy's face had been "familiar" would have been true. The guy probably could have gotten through a lie detector test.

Unfortunately, Richard had lost the ability to interrogate Minoth. Only days earlier, Minoth had called the Supervisor with tragic news. His mother had called to say that his father had had a heart attack, and he needed to go home immediately. No one had seen him since, and Richard was becoming convinced that John Minoth had never existed, in spite of the fact that his identity when hired had been checked pretty thoroughly. The man even had a social security number that matched the name John Minoth, but as it turned out the number seemingly belonged to a librarian by that name who lived in Maine.

In spite of the disappointment of losing Minoth, his disappearance provided a piece of circumstantial evidence that Richard had to consider: if Minoth had been a plant, Kline's death had been planned weeks prior to when it was carried out.

Richard picked up his pen and looked down at his "Names" page. He drew another line from the bubble at the top and connected it to an empty rectangle. At the top of the rectangle he wrote "Homicide Plan." In the rectangle, he wrote,
Pre-meditated, weeks
$$$$$

* * *

Linda Frederick looked perfect as she walked through the halls of power in Washington, D.C. After getting over her failure to recognize Huxley's tactic, she had mentally re-grouped. She was now prepared to be a general in this war. She was calm, cool, collected. The next Committee Meeting was going to go so well that she was looking forward to it.

She had called Huxley, who did not answer again. But she left a message, apologizing for sounding so upset and, in the same breath, justifying her actions inside a comment about how brilliant his plan was.

She arrived in the office of the sympathetic moderate from the other day, where she was expected.

"Thanks for trusting me the other day. I'm sorry I didn't have a chance to get to you before the meeting to explain the plan."

"You're welcome. But now, I really need to know. I'm taking all kinds of flack for supposedly ignoring evidence and going against statements I made during my last campaign."

Linda handed him a manila folder containing a copy of the opposition's article which had helped her understand what Huxley was doing. "This summarizes it pretty well. Unfortunately, as it shows, some of them have figured it out already. Bottom line: we can trust Huxley. He was right to try to push this thing into the courts because, as I'm sure you've noticed, that's where our power base is right now. All we have to do is stall on the legislative end. The courts will rule in our favor, and then the public outcry for us to do something will fade."

"Fade, yes, but it's not going to go away. If this research is really accurate, it's going to come back again in a few years, and it will be impossible to sweep under the rug then."

"Probably. But in the meantime, everyone will continue to get paid as usual. Our job is to keep the people who elected us happy. A few years from now, we may not be here, then it will be someone else's headache. If we are still here, then we still have the perfect justification for our decision: we followed what the court said. If someone accuses us of having made the wrong decision, we just point to the black robes and remind everyone that it was their call; we just did what we were told."

"There are a few out there who regularly accuse me of letting the black robes do my job."

"But obviously, they are too few to get someone else elected. Obviously, the majority of voters are comfortable with how we are making things happen, or we wouldn't be here." Frederick smiled in a way that looked almost sinister, and raised her eyebrows. "So you see, we will never be accused of doing anything wrong – we are simply doing the will of those who elected us. If blame is ever laid anywhere, it will be at their own feet."

* * *

Richard was walking around the yard where he had found the clean fill. Today, he was once again playing the part of Christopher Mills. He was confident that no one was around, so he began talking to himself.

"OK. Minoth approaches Mills, offers to pay him for a job that involves bringing the truck here. Minoth gets Mills to the yard, and Mills brings the truck here, as planned."

So far, the idea was very feasible. Richard had checked on the ownership of this land. It had been purchased about the same time Minoth was hired at the yard by a medical doctor from the city who liked to hunt. What intrigued Richard was the fact that the land also included the cabin where Mills' body had been found. The sale looked real clean – but it seemed too coincidental. The land had not been for sale. The medical doctor's land agent had approached the previous owner with a nice deal, saying that his client was interested in the property. A friend of a friend at the hospital had mentioned this nice little cabin situated on 50 acres, and the doctor was looking for just such a place. Richard had tried to follow the "friend of a friend" trail, but it dead-ended in a supposed conversation in a break room where someone had given someone a name on a piece of paper, and who exactly had written the name was forgotten.

This little area where Richard was walking appeared to have been a tentative building site for a main house. The medical doctor said that he had been told about it, and it pleased him. He liked the area, and it was very possible that he might retire to it. Richard looked around. The doctor had spoken as though the site may have been cleared by the previous owner. But when Richard went over his exact words, he realized that the doctor said only that he had seen the possible building site. He did not say that it had been cleared; nor did he say that he had cleared it. Richard had spoken to the previous owner and knew that it had been cleared since the sale.

It was all circumstantial, but too perfect to ignore. The clincher was the doctor's specialty. When Richard asked, he had replied that he was an Ob-Gyn. Richard was smart enough to know what this might mean, and decided to ask a very direct question. "So do you perform abortions?"

The physician had not liked being asked the question, but did a good job of covering this fact. His reply had been simply, "Only when necessary." The expression plus the answer had told Richard a lot.

* * *

Jacob Davidson was beyond exhausted. He knew he had to sleep. He could not do good research when he was this tired. The problem was, for the past week, whenever he had tried to lie down to sleep, his mind had continued chewing on the problem, and then sleep would elude him. Thankfully, not any more. One more piece of the puzzle had fallen into place just a few hours earlier. Now he was home. He had exercised, eaten a good meal, and relaxed to the best of his ability for an hour. Within 90 seconds of his head hitting the pillow, he was sound asleep.

*   *   *

"How's Kayla doing?"

"Great. Look at this." The tiny patient's chart was handed from one hospital team member to another.

"What do you think about taking her off the respirator?"

"I think she might handle it well. C-Pap would be much more comfortable."

"Relatively speaking, of course. I shudder sometimes at what we refer to as 'comfortable.'"

"I know what you mean."

"I agree, though, that now is a good time to begin weaning her off the machine."

The young doctor who wanted so desperately for this little girl to live now stood beside her. She was the only one within earshot, and he spoke softly to her, and his voice was tinged with excitement. "You hear that, Kayla? We're going to take that annoying tube out of your throat! You'll still have help breathing; we won't ask you to do it all by yourself – I know, you're still very tiny. But you're doing well enough now that you don't need to rely totally on a machine. You're being such a good patient. I'm proud of you!" He then laid his hand gently on the top of her head. Within a few minutes, it was apparent that she appreciated the time he was giving her. A sound sleep had replaced her earlier restless behavior.

13

Richard came home and found Susan lying on the floor. She had called his cell phone while he was walking around the vacant lot, and asked him to come home a bit early, which he had. Now she was conscious, but not totally coherent. Two and a half hours later, Susan's doctor sat down to consult with him. His expression was grim.

"I'm sorry, Richard. There's nothing left to do."

Richard Serrap had known for a couple of years that the day would come when he heard those words. He had hoped that the knowledge would help him be prepared to hear them. He realized now that he was, indeed, prepared to hear them, but preparation could do nothing to lessen their impact. He looked squarely at the doctor, who had become a friend to both Susan and himself over the last four years. "Thank you, Jim, for being a good doctor, and for always being honest with us." That was as many words as Richard's voice would carry at the moment. He rubbed his eyes with his left hand, and took a deep breath.

The physician's eyes were moist, too. There were any number of words that he could say, but he knew Richard Serrap well enough by now to know the best thing to say. "Before you leave, can we pray?" Richard nodded.

Jim took a breath, trying to control his voice. "Dear Lord, we've done all that we know how to do in order to help Susan. We know that you are powerful enough to heal her in a moment, and this is what our hearts cry out for. But Your ways are not our ways. And although this sometimes means that we must encounter suffering, we have absolute confidence in your promise that You will work all things together for good for those who know Jesus as Lord. I ask for Your continued strength, help, and comfort in the lives of Susan and Richard. In Jesus' Name. Amen."

Richard stood and shook Jim's hand. "You're the best doctor we could possibly have." And then he left to sit with his wife. He was so thankful for the long weekend ahead, so that he could just spend it with Susan.

* * *

"The goal of this Committee is supposed to be legislation, not politics!"

Linda Frederick was ruffled, but trying to hide it. "You're not going to get this thing to a vote, so you might as well give up and let us get to the next item on the agenda."

" 'Get to the next item on the agenda.' Do not pass moral responsibility. Do not collect – or try to save – any lives. Linda, you're the one who is always running back to the phrase, 'These women need a choice!' And now, you are denying them the right to choose their own life!"

"Don't twist my words, Dan." The tone was cold and accusing.

"I'm not twisting your words! I'm repeating them!" He was not yelling; his tone was only very firm. "You want women to have a 'choice' – *without all the facts.* That's not a choice, that's a manipulated outcome."

"Make your motion for another vote if you want to, but you know it's useless. You're floor..."

Dan felt beaten. He understood exactly what was unfolding in front of him, and his passion for truth collided against it head-on. He felt so helpless against it, and he knew there were a few others in the room who felt the same. But they were outnumbered. Still, he made the motion for a vote, believing that it was the right thing to do. No matter what the odds were, he had to do the right thing. On the surface, he lost once again. On the surface, it appeared Linda's words had been correct: it had been a useless motion.

* * *

"I need an RT!" The tone in the nurse's voice communicated an urgency that several people responded to all at once.

Several hours later, the young doctor stood by Kayla's bedside. She was stable now, but back on the ventilator.

"Oh, Kayla! What happened? You were doing so well!" Her tiny fingers were wrapped around the tube that once again entered her throat. His voice was soft and gentle. "Are you protesting that thing? I don't blame you. I'd want to pull it out, too. Hang in there, sweetheart. You can do it. Keep fighting. You had a setback. That's OK. Just don't give up – OK?" Perhaps in response to his tender voice, Kayla let go of the respirator tube, curled her little fist up next to her cheek, and fell asleep.

* * *

Monday had been Memorial Day, and Richard was prepared to take Tuesday off from work. But Susan was much improved, relatively speaking, and wanted him to go. "Really, honey. I'll be fine. Mom will be over this afternoon to check on me, and I'll probably sleep a lot anyway. I'm pretty tired today. I've been praying a lot for this case you're working, and I want to see you solve it."

Richard considered his wife's words as he drove. "I want to see you solve it." It was more than just a phrase of encouragement. It was a vote of confidence. Sometimes these cases took months, and Susan may not have months left. If she was going to live to see him solve it, he might only have weeks to do so.

Back to the vacant lot. Christopher Mills. It was a bit unusual, but not uncommon: the Kline death had been the official case-opener, but it was Mills' death that consumed almost all of the investigator's time for the moment. The young man was no less important, just because he was not a high-profile MD, and Richard desperately wanted to see justice done on his behalf. At the same time, Richard needed to keep reminding himself to keep Kline in the picture.

"He gets here with the truck..." Richard was talking to himself again. "...Then what? Someone's waiting for him. Someone he probably expects to see. He gets out of the truck. They kidnap him easily enough." Easily enough? Richard thought for a moment. Yes. He was a football player, a runner. He was in good enough shape to put up a fight, if he was prepared for it. But he probably would not have been prepared for it, and a gun would have taken the wind out of his ability to resist.

But wait. They couldn't risk shooting him – that would have wrecked the whole plan. It had to be a suicide bullet, and it had to be shot in the cabin. But Mills didn't know that, of course. Nevertheless, Richard considered it likely that at least two men had met and overpowered Mills – which fit his initial idea that one took Mills to the cabin while the other took the truck. On top of this, Mills "needed" to get beat up a little bit at this point – close to the same time that the accident happened. And the person inflicting the bruises had to make them consistent with a car accident, not a fist-fight.

Richard made a mental note to double-check this point with the coroner. Could the bruises have been inflicted in either manner?

*   *   *

"I went back over my initial report, Richard, and the answer is 'Yes.' Those bruises are really more consistent with an auto accident than a fist-fight. But such injuries could be inflicted with intent – if someone knew what they were doing. Also, as I reviewed this for you, something else jumped out at me. As you probably recall, in my initial report, the bruises were most consistent with the hypothesis that the victim had been the driver of the vehicle, and that he was wearing a lap belt with no shoulder restraint. This aroused a little suspicion, since it meant that, for some reason, the boy had put on the seat belt, and then slipped out of the top part of it – not a typical action by a thieving joy-rider. Your new question makes me go back to the point, because if someone wanted to fake this, perhaps they did not want to deal with the challenge of a bruise that would match a driver's-side shoulder restraint. "

"Good work, Jim." Richard mentally added yet another piece to the puzzle.

Was Mills forced to ride in the truck when it crashed into Kline's car? Was he conscious? Just before the crash, was Mills a hostage? Or had he been working all this time on the "job" that he was hired for? Possibly, Mills was a willing passenger who was riding along, thinking that they were on their way to take the truck back to the yard. But that didn't work: if he'd had his seat belt on, then he would have had a bruise from a passenger's-side shoulder strap; if he didn't have his seat belt on, he would have hit the windshield. There was a tiny chance that he had only the lap belt on, but Richard figured this very unlikely. It would have been much easier for the killers if Mills had not been in the truck at all. If he had been in the truck then, in the seconds following the crash, Mills could have been difficult to move – either because they had already rendered him unconscious or because he may have objected to running from the scene of the accident. They would have had to move very quickly because of the risk, be it ever so slim, of another car coming along. They would have had to get him into another car, away from the scene, and to the cabin, likely against his will.

Mills was probably never at the scene of the Kline homicide.

* * *

Jacob Davidson looked up to see one his colleagues walk into the room with a telling smile on her face. He could not keep the excitement out of his voice. "Success?"

"Success."

They were one step closer to completing this rushed phase of the research.

* * *

"Hi, Miss Kayla. How are you doing today?"

"Very well." It was the nurse's voice who had answered, and Dr. Jones turned towards her.

"I'd like to try to take her off the respirator again, hopefully later today."

"I'm sure Kayla will like that," she replied, really meaning it. This tiny child had become special to everyone in the Unit.

14

By now, the initial buzz and frenzied excitement with which the media treated Huxley's press release had died down. For a while, it had been included in every newscast, in one form or another. It was "news." People wrote to the editor with their opinions. Talk radio discussed it. It was a topic of conversation in lunch rooms across the nation. Key law suits were filed.

But then, as Huxley had counted on, it became "old news." It did not stop all at once; people would not forget it so easily. But the media knew how to fizzle something out slowly, on just the right time schedule, to make it appear that its lack of publicity was due to its lack of relevance.

The court cases were still alive, of course. And a few "radicals" continued to carry the standard of the new research, trying to prevent the public from forgetting what they were so convicted about only a week before. But all-in-all, Huxley was pleased. As long as most people stopped thinking about it, most of the pressure would come off the legislators. This would allow Linda and the others the stall time they needed while the courts provided the best temporary solution possible.

Carl Huxley sat in his living room, sipping a glass of wine. For the first time, he was enjoying having this job. Carl's wife had noticed that the initial stress of his new job was finally waning, and she was thankful.

* * *

"You did a good job in the Committee Meeting."
"Thank you."

A single piece of paper changed hands. $30 million had just been transferred.

*   *   *

Linda Frederick sat in her own living room, also sipping a glass of wine. It was a heady feeling to possess $30 million. Well, $10 million, actually; the other $20 would find its way into the hands of others. Still, $10 million was a nice number. It would probably be several years before she could spend any of it, and this chafed a little. But at this point, the initial thrill more than counterbalanced the disappointment.

*   *   *

Jacob Davidson spent ten minutes briefing his team on the progress they had made so far. He wanted to commend them for their hard work, and he was incredibly pleased with their progress. He acknowledged the sacrifices that many of them were making for the sake of keeping this research alive. And then he dismissed them.

He had not yet received any subpoenas to testify in court. They would be coming soon, he was sure. But every week was precious. Time. He needed time. In spite of the tremendous progress, there was still an Intensive-Care level of threat posed to this project. It probably could not survive in court yet. Nevertheless, it was gaining strength every day. He was optimistic that, by the time he laid his hand on the Bible in the courtroom, they would have what they needed to win. The evidence might be a little underweight, compared to ideal maturity, but it would survive just fine.

Davidson was not sharing any of his optimism with Huxley, and had admonished his team to keep silent on their progress. They understood too well. At this point, the best thing they could do would be to surprise the courts (and everyone else) with how much they had accomplished in so short a time.

* * *

"How'd that procedure go?"

"Good. The transition was tricky for a moment, but now she's back on C-Pap and liking it. She's breathing so well that you'd think she's determined to never be on another respirator ever again."

The physician looked at the sleeping infant. "That's a good goal, Kayla."

* * *

Richard looked around the vacant lot yet again. Mills had brought the truck here. He had been overpowered. The truck got a new driver. It waited here a few hours. Then it went to sit at the pull-off where the lookout would call to say it was time to go meet Dr. Kline – head-on. In the meantime, Mills' captors took him to the cabin. How? They may have drugged him. It was a few days before he was killed, so they could have given him something that would not be found in an autopsy. But then it would have been a bit difficult to get him to the cabin. Plus, they would have had to keep him restrained there. It was certainly possible, but...

But did they have to overpower him? Richard suddenly realized that, at the point Mills handed over the truck, he knew of nothing out of the ordinary. It was possible that they had convinced him one way or another to go to the cabin voluntarily. A nagging thought worked its way into Richard's mind. He did not give it much weight at first, but it did not go away. Eventually, Richard followed it down a rabbit trail that he decided was worth pursuing.

Richard made a couple of calls on his cell phone, and set up an appointment for later that day.

* * *

Now Richard sat across an informal table from Mills' girlfriend. She appeared nervous that Richard had wanted to talk to her again.

"Amelia, the first time we spoke, you indicated that Christopher had planned a special get-together for just the two of you the weekend after the accident. Do you remember where he had planned to take you?"

Amelia was nervous. She was pretty sure that she had said only that Christopher had planned "a surprise." But this cop knew they had planned to go somewhere? Had she been so nervous that she had let it slip? She was more nervous now than she had been then, and she couldn't think past the noise of her heart pounding in her ears.

"I'm trying to remember." That was the truth, sort of. But she wasn't trying to remember the answer to his question, but to her own.

Richard could read the girl's nervousness, and followed the hunch that had gotten him this far. "Amelia, I'm going to tell you something important about Christopher's death, but you have to promise not to tell anyone, not even Christopher's parents, for now."

This brought Amelia's head up, which the detective knew was a good sign.

"Christopher may have been tricked into taking that truck. I know you wanted to protect him before. I can understand that. But now you have to tell me everything that you can remember, because I need help in order to try to clear Christopher's name."

Amelia looked at the detective who was so convincing. He seemed to be a nice guy. He seemed to really care. But didn't all good detectives act this way – do whatever it took to gain the confidence of the person they wanted to get information out of? Amelia had had enough of a rough childhood to prevent her from trusting anyone quite so easily. But the knowledge that came from being street smart had its advantages. Richard wasn't the only one of the pair who had learned how to read a lie on someone's face.

"Are you telling me the truth?"

Richard looked her square in the eye. "Yes."

Amelia was still incredibly nervous. She still feared that this was a trick. But Detective Serrap obviously knew that she knew more than she was telling. If she tried to lie, it might get her in serious trouble. Most importantly, Amelia believed that he was telling her the truth, and more than anything, she wanted to help clear Chris' name. "A week before the accident, Chris said that he had been offered a special job. He wasn't supposed to tell *anyone*. Some guy who worked for the county needed someone to drive a dump truck. He wasn't supposed to hire a driver without going through all kinds of red tape with the County, but he said he'd heard that Chris was capable, and could use the money. It was *$1000* – for just a few hours' of work. He told Chris that he could get in a lot of trouble for bypassing the rules, so if Chris wanted to do it, he had to be really quiet about it because they'd both be breaking the law. I tried to talk him out of it, but he had all these plans, and wouldn't listen. After the accident, I knew that the story about Chris stealing the truck wasn't true. But the truth was worse, it seemed. If everyone thought he had stolen it just for fun, then that doctor's death was really accidental. But if everyone knew that Chris had planned to do something illegal to begin with, then the blame would be even worse." By now, Amelia was close to crying, and almost unable to speak. "For the sake of Chris' memory, I just couldn't tell what I knew – I *couldn't!*" And then she broke down and cried.

Richard passed her some Kleenex, and waited patiently while the storm of emotion passed.

"I sort of wanted to tell, 'cause the guy at the county was partly to blame, too. But I couldn't – not without making Chris look like a criminal, too. I decided that protecting Chris' memory was more important than getting the other guy in trouble. It wasn't going to bring the dead doctor back, either way."

Richard let the silence run on a bit, giving Amelia time to think, time to remember anything else. Then he reassured her. "Amelia, you're doing the right thing." That brought a fresh wave of tears, and more Kleenex. Richard waited. When it seemed that she was done speaking, and had her emotions somewhat under control, Richard followed the trail farther.

"Amelia, did Chris say anything about a cabin?" The location of the dead body had not been released to the public.

Amelia thought hard. "I don't think so."

"Did you guys plan to go away for the weekend?"

"Well, sort of. Chris said that he was going to take me someplace really special after he got paid. I wasn't sure how long he was planning to be out, and I didn't want to ask. I wanted to be surprised. So I arranged to stay over at a friend's house Saturday night, and the friend promised to cover for me if I never got there."

"Did you and Chris ever go camping?"

"No, but I know he loved it."

"If he had taken you out to a little secluded cabin in the middle of the woods, would that have been 'someplace really special' to you?"

Amelia's eyes were downcast, and her emotions were near the surface of her voice. "Yeah." She looked up at Richard, the tears in her eyes threatening once again to spill over. "Is that what he was planning?"

"I don't know, Amelia. I wish I did." Richard decided not to tell her any more. He'd learned long ago that, when in doubt, it was better not to give any more information. Additional information could always be shared later; it could never be taken back. "I know there are still a lot of pieces missing, and once I have a complete picture, I will make sure you get a chance to see it. I want you to know, though, that you have helped *a lot*."

Richard went over what she had told him, making sure that nothing major changed the second time. He knew that the girl was dealing with guilt, and had enough training to walk her through an initial counseling session. She agreed to speak with one of the Department's counselors, for which he was thankful. She needed a bit of help, and the Department counselors could provide it without compromising the fact that the information needed to be kept confidential.

* * *

After Amelia left, Richard went over every word of the interview. This had been one of those bitter-sweet moments that investigations often held. Internally, he reacted almost joyously as Amelia spoke, for her words answered a lot of questions. But externally, he could not appear joyous to the young woman, who was still deeply grieving and dealing with guilt.

His suspicions about missing County employee John Minoth were proving true.

15

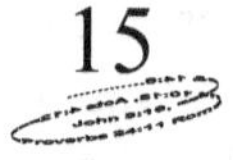

"Your people are doing well."

"Thank you."

"But they need to do better."

Silence. Xenan wore $10,000 suits, handed out $10 million checks to Congressional employees, and oversaw more than $10 billion of the organization's assets. Only one person in the world, literally, would ever dare berate him.

"You're in charge of the United States. Up until now, you have done very admirably. But don't forget – we are a world-wide organization, and you are just a piece. Of course, you are a very important piece – perhaps the single most important piece: which is why this new development must be swiftly dealt with. If you fail in America, I need not tell you what will happen in all our other countries."

No more words were spoken at this point, but a wealth of communication passed back and forth across the table. These were men so intimately associated with the depths of deceptive power that words were often a mere formality.

The meeting ended when the Superior rose and left, leaving Xenan alone at the table.

This was only the second time he had ever been berated, and a hot determination rose in him: it *would* be the last. He was exceptionally skilled at pouring all his anger into a goal, and this goal was a no-compromise imperative. For years, Xenan had known that he should be the Superior's successor, but the when and the how were elusive. Now it was clear. His greatest liability – this new line of research – would become his greatest asset.

Successfully dealing with Jacobson's research would secure his position as the next Superior.

The longer Xenan sat there, the more spiritually drunk he became on the opportunity. He had all the resources he could possibly need. More importantly, he had already agreed to pay any personal price: this, more than anything else, had gotten him this far, and it would be the key to gaining his ultimate goal. Even he did not know what the price would be, but he chose not to care; he would gain the whole world.

*   *   *

The vacant lot. Richard came here once again, prepared to move to the next stage of the investigation. The only thing he had left to do here was try to figure out the path to the cabin. Knowing that he was standing on the same legal lot that the cabin sat on, he wondered if there was any sort of a path that connected the cabin with this logical site for a main house. Map and GPS in hand, he checked out a few hunches, but found nothing. No path had been visible from the air, but the trees were so thick that Richard wanted to eye-ball it. He couldn't find a thing.

The worn path to the cabin from the road – the one he had walked before – appeared to be the only obvious way. He took his car down to that spot, parked, and began the hike. Once at the cabin, he scouted around, trying to find a path back towards the vacant lot. There wasn't one – not that he could find, at least.

Richard conceded that the worn path was probably the way Mills came – or had been brought. Now he hiked it again, back to the car. Then back to the cabin. Each section of the trail was scrutinized. It was mostly a flat walk through a pleasant wood. Pine, cedar, a few hardwoods. Richard recognized a lone dogwood tree at one point. There were one or two spots where the trail went up or down a little hill. There were some stretches where a running escapee might try to break off the trail and hide, if his pursuers were a few hundred feet behind him. There was a hunter's deer stand up in one tree which, Richard estimated, was more than 15 years old.

Richard couldn't find anything dropped on the trail. There was no evidence of a person who had tried to elude capture. Richard suspected that Mills had come along this trail willingly. Arriving at the cabin once again, he sat down on the front step. Possible scenario: Minoth makes a trip to the cabin part of the planned scenario. He offers the use of it to Mills for the weekend – for himself, for him and his girlfriend, for him and his friends, whatever. He tells Mills to take the truck to the lot. Someone will meet him there, show him the way to the cabin, and then take him back to town. Or maybe Mills was supposed to take the truck back later. Mills gets to the cabin. Maybe he is overpowered right away. Maybe the plan was that he would stay here for the weekend. If he was planning on staying here, he had not told his parents. But this would not have been the first time that he disappeared on his parents for a day or two. Forcibly or willingly, Mills remains in the cabin a few days. Then he is shot; the forged suicide note is left.

It was a great scenario, but Richard knew the evidence was still sorely lacking. Perhaps the worst part was that he still did not have a good lead on who the money brokers were. And without that, he did not have a prime suspect. If Susan was going to see him finish this case, he needed a breakthrough lead – soon.

* * *

"Hi Kayla." The young neonatologist spoke softly to the tiny girl. "You're growing so well. Look at you – you've gained eight whole ounces since you were born!

I'm proud of you." Gently, he placed his hand on her head. In less than a minute, her tiny arm reached up, so that her fingers brushed his wrist. The action captured the doctor's heart. He loved his job, but moments like this were extra special.

The doctor spent about 30 minutes just standing there with the tiny infant, providing comfort, love and encouragement by leaving his hand on her head. He never moved it; preemies liked it better if you held your hand still. Touch was so important to them, and since little Kayla did not have any family to show her love, the staff made a special effort to do so. And she was responding very well.

*   *   *

"Sue, Mom made chicken soup and fresh bread for dinner." Richard's voice was filled with gratitude. He tried to infuse encouragement into it, too. Susan was not looking well. "Can you come down and have a bite with me?"

Susan tried as hard as she could to make herself sound better than she felt, but her voice betrayed a deep weariness. "Of course." She was wearing a comfortable lounge outfit that doubled as pajamas. The fact that she had left it on all day gave Richard some insight into how poorly she was doing. Richard retrieved her robe and held it for her while she put it on it. Then they slowly started towards the kitchen.

"I love Mom's chicken soup. Ever since I was a little girl, she would make it when I didn't feel good. And it always helped me feel a little better."

For what felt like the ten thousandth time, Richard admired his wife.

*   *   *

Linda Frederick was very pleased to receive the invitation to meet Xenan. Of all the power meetings she had ever had, those involving this elusive man had been the most exhilarating. As she chose her wardrobe for the evening, she scowled for a moment. The "invitation" was actually a summons. Despite the fact that she really wanted to attend this meeting, it irked her that she had no choice. The irony between this and the fact that this man was the most powerful abortion broker in the country eluded her thinking.

She may not have much freedom with regard to attending this meeting, but she *would* choose what to wear, and Linda was giving a great deal of thought to this decision. She had no idea how many opportunities life would afford to interact with this man, and she intended to make the most of every one – especially the ones that were his idea. It may be a summons, but it was worded as a dinner invitation, and that was an exploitable loophole: it allowed her to lay aside the business suit and choose a medium-red formal dinner dress.

When they sat down to eat a couple of hours later, Linda was pleased to note that the dress made an impression, albeit un-worded. As soon as they were seated, Xenan opened the conversation.

"Huxley's plan is fine for the short-term, but we must have a long-term method for managing this research. Otherwise, it will seriously jeopardize our goals in the coming years. I'm laying out a plan now. When your role is fully defined, I will inform you. But I wanted to remind you that your actions now ought to correspond to reality as we desire it twenty years from now; do not allow a short-term-fix mentality to guide your thinking."

Linda understood what he was saying, but it did not seem realistic at all to her. "You are implying that this 'problem' can be *permanently* buried?"

"Of course."

Those words clashed sharply with the facts, as Linda knew them. And yet, it was more than just confidence that supported their weight. The way this man said them, and the expression on his face, indicated some sort of evil prophetic ability.

A warning fear surfaced in Linda's conscience. This man was introducing her to a reality other than the one she was aware of. Reality, as she understood it ten seconds ago, demanded that Davidson's line of research would eventually come of age and impact the abortion industry; the best that they could do would be to delay its fruition. Now she was being introduced to a reality in which the problematic research "went away" entirely, never having the opportunity to compromise the industry. The fear came from Linda's realization that this man had not only the intention, but the apparent ability, to force his version of reality into history.

Linda Frederick, a champion of choice, now had a very important one: heed the fear, and flee, or allow the depths of power that she had just glimpsed to fascinate and intoxicate her. A few seconds passed. Then she raised her wine glass, as if in toast to his words. The glasses clinked, they each drank a sip, and Linda Frederick tasted the ability to communicate across a table, wordlessly. This depth of power was not only intoxicating, but addicting.

* * *

"What's wrong?"

"I wish I knew. We haven't done a thing differently the past several days, have we? Have I missed something on her chart?"

The physician looked through the pages, reading numbers and comments. "Failure to Thrive."

The medical student was familiar with the phrase, but had never come across it before. His expression begged for an elaboration.

"Doctor Jones was making a point of spending at least 30 minutes twice a day with Kayla, either talking to her or holding her head. He's been out with a cold for a week now, and others have been understandably busy, because he's not the only one out sick. She's missing the love."

The young student acknowledged the prognosis. He had never realized that simple touch could have such a profound impact on a person this tiny. He had thought it better not to disturb her, because it looked like she was sleeping. Now he wondered if that "sleep" had not been more like a depression-induced lethargy.

* * *

Richard knocked on the open door, and when the man at the desk looked up, he walked in.

"John, I need something from what we know about Frederick as a pointer."

John shook his head and held his palms out, face-up. He clearly wished that he could give Richard exactly what he wanted, but the well seemed dry. "Do you want me to go over the transcripts with you?"

John had been diligent to make sure Richard received copies of every transcript as soon as they were available, and Richard had read them. No lead that could be followed had jumped out at either man, but at this point, it was worth going through them again.

"Sure. Let's start at the beginning..."

Meanwhile, Susan Serrap prayed for her husband, and for the members of his team.

* * *

Judith Kline walked out to the mailbox and retrieved a bundle of mail. It was a beautiful afternoon, cool and crisp. She had enjoyed this routine for decades – walking to the end of the driveway each day. It wasn't far, but it was something she always looked forward to, especially if the sun was shining. She was still grieving, but thankfully she was to the place now where little activities like this were enjoyable again.

Mrs. Kline placed the mail into the little canvas bag that she brought for this purpose, wondering for the umpteenth time why she didn't put a garbage can right behind the mailbox. If she sorted the mail here, then two-thirds of it could go right into the trash can. Then she wouldn't have to carry it all the way back to the house, and then her son wouldn't have to wheel it all the way back to the curb on the day the garbage truck came. Judith smiled and shook her head at the idea, but it also brought a tinge of heartache. Her son wheeled the garbage now, because his dad wasn't around to do it anymore.

Back at the house, Judith sat down at the kitchen table and began sorting the mail. Then the phone rang. It was their daughter-in-law. They would love it if she could come over for dinner tonight. Nothing special, but it had been more than a week since her last visit. Was she up to it?

"Thanks, Nancy. I would love it. What time would you like me to come?"

"We'll eat at 5:30, but the kids would love it if Grandma came right now, and got in a little play time before dinner. That's up to you, though."

"I'll be there in about 15 minutes."

"Great. See you then."

Judith Kline left the unsorted mail on the table and got ready to go play with her grandchildren. They were each a well-behaved blessing, and God never failed to minister to her through them. If she had noticed the bulky envelope with the unusual return address, she would have opened it. But it never caught her eye.

*   *   *

In another part of the country, David Jacobson was sorting through his mail. There was only one piece that he was going to bother to open. The rest got thrown away, and then he sat down at the table with an unusual-looking white envelope. It was hand-addressed, and the return address on it, also hand-written, was a sheriff's office somewhere back east. He stared at the front of the envelope a moment longer and scowled. He thought he recognized the handwriting, but... Oh well, might as well open it.

Inside, David found a key and a letter. One glance at the name of the letter's author, and he was glad he was sitting down. If anyone had been watching, they would have noticed that, suddenly, the surrounding world did not exist for David Jacobson; every ounce of his attention and concentration was placed on the letter in front of him. He read it a few times, then looked at the key on the table for several seconds. He laid the paper down, took a deep breath, and let it out. He put his elbows on the table, bowed his head, and rested it on his hands. For several moments, he just sat in that position, trying to process what he had just learned. Finally, he spoke. "Lord, please help everybody to do the right thing." For now, that was all he could pray. But it was enough. Those nine words, plus the continuous cries of his heart, would not be ignored.

He probably was not going to accomplish a lot of research today; that was OK. This was more important. David reached for his cell phone and dialed a number. It took a bit of explaining, three phone transfers, a total of 15 minutes on hold, and some patience, but eventually the voice on the other end of the line was that of Detective Richard Serrap.

*   *   *

"Hi, Kayla. I missed you! But it was better for you that I stayed home for a few days. I had an awful cold. Since I'm a grown-up, my body could fight it easily, and now I'm all over it. But I couldn't risk giving it to you." Dr. Gary Jones was speaking in a soft, gentle voice, as if he were reading "The Three Little Pigs." Kayla apparently loved it. Her breathing became stronger and deeper. He kept talking in the same tone. "Some of these viruses could kill you, you see. You're so tiny, and fragile. We need to protect you a little longer. But you're doing so well! Look at you! You're growing, and breathing well. Pretty soon you won't need our special protection at all. You'll be able to go to a home where a Mom and Dad love you, and will take good care of you. You might even have older brothers and sisters. You'd like that, 'cause they would help watch over you."

Gary gently laid his hand on the top of Kayla's head. She was lying on her stomach. A few moments later, she reached her tiny hand up beside her head, so that her fingers, which were now almost an eighth of an inch wide, brushed his arm. Gary thought about what he had told Kayla. Oh, how he longed to see her in a good, loving home. It grieved his heart that he did not have more power in the process of choosing her home. But he was a wise man, and he would not allow that grief to overwhelm or depress him. For the next 30 minutes, Dr. Gary stood beside Kayla, his hand on her head. He silently petitioned the Heavenly Father of this little orphan, asking for a Godly home for her. The whole time, her fingers rested against his arm. It was an emotional picture for the doctor. It seemed as if Kayla were reaching up to him, touching him, so that she might join the request that he was making on her behalf. He knew that this wasn't "really" the case, but he marveled at how God took the child's ability to reach out for what she knew she needed – physical touch – and used it to draw a picture of what she did not yet understand that she needed – her Heavenly Father's help.

16

Xenan sat in a semi-dark room. It was lavishly decorated and furnished, but an oppressive dreariness hung over it. His eyes were open, but he was not looking at anything. He was deep in thoughtful meditation, calculating how to carry out his desires. Most of the plan was visible, but there were a few important details yet to be seen.

* * *

Judith Kline returned home, after dinner with her family, in good spirits. She parked the car in the garage – a job that Edward had always done – and walked into the house from the attached garage. It had been a good evening, and now she was pretty tired. She would get ready for bed, read for a little while, and fall asleep. She never gave the mail on the table a thought.

* * *

Linda Frederick decided that it was necessary to spend a little of her $10 million in order to move gracefully within the circle of power to which she had recently been admitted. Besides, she had not gone shopping for fun in a long time. As long as she did not transfer too much at any one time, and as long as it did not sit in an account for long before it was spent, no red flags should be raised, and no one would notice. She was right, too – no one would have noticed, except that her accounts were already being watched.

* * *

"How's it going, Rich?" Susan's face was haggard-looking, Richard noticed. But her eyes were bright. As soon as he had greeted her that evening, her countenance lifted, because she read in his face the fact that something had gone very well at work. She had asked the question with almost a mischievous tone that communicated her observation that he had good news. She wanted to reassure him that she still noticed that particular expression, loved him, and cared.

Richard was joyous, but simultaneously his throat constricted and he had to swallow the emotion that rose – how he loved this woman who loved him! To see her wasting away before his very eyes, and at the same time receiving her playful teasing which emphasized her love for him... He swallowed once more as he hugged her, which he always did when he got home from work, and deliberately focused his thoughts onto the case and phone call. By the time Sue saw his face a few seconds later, the joy had overshadowed the mourning.

Susan had caught the play of emotions in Richard's face. This was a part of dealing with the disease, and it was not easy. But it had made their marriage so strong, and for this result, she was thankful. Once Richard's smile was firmly in place, he began to share some exciting, encouraging news. How she loved that smile! She was paying close attention to his words, but she was also cherishing his expression. This was the expression that Richard the Detective wore when she was allowed to glimpse the specific ways that God answered her prayers for his work.

* * *

The next morning, Judith Kline went into her kitchen and made her morning cup of coffee, as usual. She noticed the mail when she sat down at the table to do her morning Bible study. The pile got pushed aside, temporarily. It was about an hour later that she turned her attention to it.

Junk mail. Junk mail. Junk mail... And then she froze. Hand shaking slightly, she picked up the next envelope. The return address, hand-written, was the local police station, but the handwriting on the front was her husband's.

* * *

"It doesn't rain but what it pours." Ken was standing beside Richard's desk, looking over the evidence that lay spread out.

"I'll take this kind of monsoon any day!" Richard had already been through each piece of paper at least twice, but he was at least a couple hours away from being finished with all the material.

"Walk me through it, will you? Then I'll be able to get started on your request."

"First, I received a call from David Jacobson – the researcher in charge of the abortion-breast cancer link. He had received an envelope in the mail, with this letter in it:

David,

If you are reading this letter, then I am most likely dead. What a way to begin! Yet, my concern for my life and your research drives me to make plans for such an eventuality. Let me start at the beginning...

When I first saw your research, my heart rejoiced. It is powerful and solid, and I felt privileged to be sitting in this Chair at that time. I wanted your work to receive the attention that it deserves. It was not on my desk twenty-four hours, though, before I was faced with the forces that oppose the truth. I don't know how, but Linda Frederick – a staunch abortion advocate in the House – sees inside this office as though it had glass walls. Past events involving her tipped me off: as soon as I finished reading your initial report, I knew that she would be calling. Sure enough, the very day that I read your full report, she contacted me.

For more than two years, I wanted to know who was behind her phone calls and knowledge. I wanted to know who was using her to try to influence this Chair. Not that I could be influenced – I give you my word, I never was. But I wanted to try to gather some knowledge, some evidence, that could be used to make it stop. I had the distinct impression that someone very powerful – who opposed truth – was very interested in manipulating this Chair. They could not touch me, but I wanted to make sure the issue was dealt with before someone else was sitting here.

I was nearly certain that Frederick was going to call, and so I decided to use the opportunity – there would never be a better one. I had a hunch about who was using her, and the best way to test it was with a dollar amount. When she called, I offered to 're-interpret' the evidence in exchange for $500 million. In a later phone call, she countered with $450 million, which I refused. That told me what I needed to know. There is only one abortion provider I know of who would be willing and able to pay me that much, that quickly. We have discussed them often. Suffice it to say, it is not surprising that they lust after this Chair.

Besides wanting to verify who the powerbroker behind Frederick is, I had another motive for playing the big-money game: I need to buy time. You were right in handing this off to me when you did. However, the fact that it has leaked out of this office so soon has placed it in critical danger. And when I try to imagine how desperately they want this thing killed, I conclude that they are likely to go to great lengths to achieve their ends.

If something were to happen to me, Carl Huxley would get my Chair. When I combine this with the adversary's power, money, motivation to retain their power and money, desperation to hide the truth, and lack of respect for human life, I can not help but wonder how much longer I will be alive. By playing the money game, I am hoping that I am buying the time I need to help your research attain maturity.

You know me well enough to know that I would never take their money or compromise your research. I wanted to give you some of those details, though, because there is a slight possibility that Frederick taped our phone calls. If she did, then the day may come when they are used to paint a wrongful picture of my intentions. If this ever happens, please make sure that Judy and the rest of my family know the truth.

If you have received this envelope, then Judy has received one that is similar. Please speak with her. I am concerned that one of you may be with me if I am killed, and that is one of the reasons that you each have your own copy.

I am placing the return address as the police because, should something happen to both of you, then this information will hopefully be placed in the hands of those who need it.

I have made a complete copy of the research report you gave me, Dave. It is in a safety-deposit box, which the enclosed key will open. All the information you need about the bank and the box is attached. In that box you will also find a sworn affidavit which details everything which I think might be considered useful evidence in a court of law. I've done enough research on our infamous abortion provider that their ability and motive to carry out my murder will be obvious. This is not to say that I am certain of who ought to be convicted, if something has happened to me. It's just that money talks, and I only know of one member of that industry who can speak a language with such high dollar amounts.

Now I must warn you: be careful. I am their immediate threat, but you are their long-term nemesis. At this point, I don't see how they could keep this suppressed forever, even with all their money and power. But I suspect they will try. The top echelon sees itself as untouchable by the law, which gives them the idea that they have nothing to lose by trying, and much to lose if they don't.

I look forward to seeing you in Heaven, but I'd really like to see you live a long, fruitful life on earth first. Keep up the great research. Remember, there's a whole cloud of us cheering you on.

May God bless you & keep you,

Dr. Edward Kline

Richard turned to Ken. "I spoke on the phone with Jacobson for quite a while. That was a couple of days ago. He over-nighted me the whole package, which is this pile. It arrived yesterday. Also, yesterday, I got a call from Judith Kline. She received a letter nearly identical to Jacobson's."

"But how did they get them? Do we know how Dr. Kline arranged these posthumous mailings?"

"Well, sort of. These were in Judy's envelope." Richard pointed to a little stack that included a note, a $50 bill, and a very weather-beaten, bubble-padded envelope. "As near as I can figure, the scenario went something like this: Dr. Kline wrote the letters to Jacobson and his wife, put the keys in the envelopes, and sealed them. Then he took a $50 bill and attached it to this note." Richard pointed to a little post-it note which read, "To whoever finds this envelope – Would you please put the two enclosed envelopes in the mail. The postage is all paid and attached. Keep the $50 for your trouble. Thank you."

"Then Kline put his two pre-addressed, postage-paid envelopes in this larger envelope, along with the money and the note. The large envelope he then, apparently, left alongside State Route 30. Don't ask me why he picked that spot.

"Then, a couple of days ago, someone found the big envelope. They wrote on it," Richard pointed, "'Found in the brush alongside State Route 30, north-bound side, about thirty feet south of the bridge over Deer Creek.' They opened it, found the money and the note of instruction, and sort-of complied. They cut open Judith Kline's letter – see? – and put the large envelope, the money, and Dr. Kline's little post-it note in with her letter. Then they taped it closed, and mailed it, along with Jacobson's."

Ken's brow wrinkled. "But then postage on Judith Kline's letter would not have been enough."

Richard smiled. "I was getting there. It appears that the original postage on her letter was added to. These stamps are the same as on Jacobson's. But then there's this added postage. My guess is that the person who found the big envelope wanted us to know as much as possible, so they paid a couple dollars out of their pocket in order to add the additional envelope, note, and money."

"Any idea who found them?"

"Not really. There's the handwriting on the big envelope, but nobody to match it against. The letters were canceled in the local post office. I've interviewed everyone who could have possibly sold that extra postage or seen the envelopes, but no one remembers them. I don't think the sender brought them in the building. The extra postage that was paid was actually too much. I think whoever it was added a bunch of their own stamps until they were sure it had enough postage, and dropped it in the outside box. It's a small enough envelope to do this, and the additional postage is a strip of 42 cent stamps – three too many, in fact."

Ken nodded. "Apparently, whoever it was did not want to be involved in a police investigation."

"And a murder investigation at that. The fact that they opened Judith's letter in order to add to the envelope means that they probably read it. I can think of a dozen reasons why someone might choose to send this stuff anonymously."

"Prints? If they read Judith's letter?"

"Poor at best. The lab did find some which are probably not Judith Kline's or Edward Kline's on that letter. So far, the computer has not come up with a match. But I don't expect it to. I'll bet our helpful citizen has never been fingerprinted, so they're not in the computer."

Ken nodded, agreeing. "So what was in the safety deposit box?"

"This pile." Richard pointed to the thickest of all the stacks. "Judith Kline readily agreed to go to the bank with me and open the box. I've gone through it all, except that I have not read the technical medical research report word-for-word. The most interesting part is Dr. Kline's research on the largest abortion provider in the world. Look at some of these numbers."

Ken looked, and although he knew the organization was big, his eyebrows went up when he saw the figures. "Is this right?!? *$2.5 billion?! ? And they take in that much every year?!?*"

"Our best Accounts Researcher already has a copy and is in the midst of corroborating what Dr. Kline found. But yes, the numbers appear accurate. Kline carefully documented every single step of his research."

Richard had never seen a stunned expression on his boss' face, but now there was no better description of Ken's countenance. His lengthy silence as he perused the numbers that Kline had so carefully compiled was broken only by a barely audible, "I can't believe it." But Richard knew by the tone of his voice that he did believe the evidence; Ken's comment was meant to reflect his disbelief that the truth about such a lucrative murder industry could be concealed from so many, so well, for so long. "I've heard of this group, of course. Who hasn't? But I had no idea that they are directly responsible for tens of millions of abortions, every year, on a world-wide basis..." Ken turned a few more pages. "...and with my tax money?"

Richard nodded, but did not say a word. His boss was listening to the evidence in his hands, and that was what Richard wanted the most. He had prayed for Ken for years, but Ken had never conceded that a Biblical worldview was worth having. He respected Richard a lot, but he could be a great cop without God having to exist. But now, something about the papers in his hands seemed to speak to Ken's heart in a way that Richard's words had never accomplished. Finally, Ken digested the data. When he looked up, his expression was professional, and Richard was wise enough to know that the conversation had to remain that way. Ken's voice betrayed none of the emotion that Richard had seen sweep the man's face moments before. "You think these people murdered Kline?"

"I do."

There were a few seconds of silence as Ken nodded his head. "I agree. I'll do everything in my power to access the information you asked for." And then he left Richard's office to do exactly that.

Richard bowed his head briefly and prayed a prayer of thanks. Then he took out his manila folder, and wrote a name – the name of the abortion provider – in the empty oval at the top of the page with the dollar sign in it. He had his suspect. Now to prove it. As he drove home that night, hope rose in his heart that Sue really might live to see him solve this case.

17

Xenan looked at the single piece of paper in front of him. At the bottom was a figure: $500 million. Such a small price. He laughed. How ironic – the total cost of his plan to ensure his position as the next Superior was the same as the amount that Edward Kline had first named as a bribe. His eyes narrowed a little when he thought of that man. Foolish man! Kline had been a thorn in the organization's side from the moment he sat down in that Chair. And he had begun to figure out too much. When he had asked for the bribe, Xenan thought for a moment that the man's inquiries into the organization's finances had been for that purpose. Xenan had relished the possibility that the good doctor was not as principled as he led everyone to believe. But the doctor's smokescreen had lasted less than 48 hours before it was obvious that such was wishful thinking. Xenan's original plan was carried out, having been delayed for only two days.

A knock on the door preceded the entrance of a polished, but cold-looking man. "You have something for me, Sir?"

Xenan handed him the paper wordlessly. The man perused it, and nodded. "I understand." Then he left.

* * *

"Where did this infection come from?!?!?" Dr. Gary Jones was upset, but not at the person in front of him. "NO! Kayla has not fought this hard, and come this far to die of infection!" Jones stopped, swallowed, and calmed down. It could be life-threatening if they weren't careful, but it was not quite as serious as his reaction had implied.

He laid out a plan of treatment. Then he walked down the hall to the baby's bedside.

Gary gently laid his comparatively giant hand on the little girl's head. Tears stung his eyes, and he blinked quickly to regain his composure. He couldn't speak for a moment. Then, in a voice that was a bare whisper, he reassured his tiny patient that he was there, and that she was going to be OK. For the next thirty minutes, he silently petitioned the best Healer he knew to touch her tiny body.

*   *   *

In the city where David Jacobson and his team were working, $200 million changed hands. In the city where Richard Serrap lived, $100 million changed hands. In the city where Linda Frederick's main office was located, $100 million changed hands. And in the bank account of one of Washington D.C.'s most powerful lobbies, $100 million began to trickle in from various sources. The total of all these transactions: $500 million.

*   *   *

Richard looked up as Ken entered his office. Ken's expression did not look happy. "The good news is that our request for the account information you need was granted. The bad news is that, twenty minutes later, a higher-court judge yanked it."

Richard let out a sigh, but kept an optimistic expression. "When I think about who we're fighting, I'm not surprised. But that was only Round One, and we've got at least six more good Rounds in us. When's the appeal?"

"Not until Monday. It never ceases to amaze me how quickly judges can make decisions on a Friday afternoon – when they want to."

* * *

Linda Frederick had been looking forward to this weekend for a while. The ability to access a bit of her $10 million had taken longer than expected, but now she was finally going to indulge in a small shopping spree. A couple items had been ordered online, but for Linda, half the fun of spending large amounts of money was doing it in front of people. Dresses, jewelry, a couple of fun items. Less than $10,000 – but not much less.

* * *

"Sue, maybe we ought to go back to the hospital."

Susan's eyes filled with tears. "Richard, no. We talked about this. I want to stay home, remember? Please?" Her voice cracked, and she reached out her hand to take his. Despite the pain, her gaze held his eyes.

Richard knelt down beside the bed where his wife was dying. He wasn't about to force her to go to the hospital, not when he knew how much she wanted to remain at home. He had only suggested it because, for a moment, he had dared to hope that they could do something to prolong her life. But no. There was no hope of that anymore; the last hospital visit had confirmed this.

Richard took Susan's withered hand, held it to his mouth, and gently kissed it. "I'll never make you go." The tears coursed down his cheeks as he choked out the next words, "Don't go, Sue. Don't go." Richard held his wife's hand to his face and sobbed. And he prayed. When he finally looked up, Susan was breathing more comfortably. He kissed her forehead. "You get some sleep, and I'll get dinner."

"Is there any of Mom's soup left?"

"A little. Would you like that?"

Susan nodded. Richard began to leave the room.

"Rich?"   Immediately he turned back to her. "Would you please rub my forehead while I fall asleep?"

Richard smiled.   It was something Susan had always loved. He sat down in the chair beside the bed, and placed his hand on the top of her head in such a way that his thumb could touch her forehead. With it, he gently caressed her forehead. Sue closed her eyes, and Richard could see her whole body relax as she drifted off to sleep. He clenched his jaw to fight the emotion; he did not want to cry and wake her. How many more days would he have this privilege?

Once he was sure that she was asleep, Richard went downstairs. Sue loved toasted cheese, so he prepared two sandwiches and put them in the toaster oven. He would cook them after Sue came down. He would probably end up eating both of them, but Sue might be able to eat half of one with her soup.

Richard then called Sue's parents to give them an up-date. Thirty minutes later when he checked on her, she was awake. He insisted on carrying her downstairs. With the addition of a fresh, salted tomato slice, he was pleased to hear Sue ask for half of one of his sandwiches.

* * *

"What's this?" The FBI agent in David Jacobson's city scowled at the paper in front of him. Tips on Sleeper Cells had to be taken seriously, and he intended to follow this one. But the source was a bit suspicious. Also, something this big should have been noticed and flagged earlier. This group seemed to have condensed overnight. Something wasn't quite right.

"What's the address?"

"9310 Redwood Drive."

The agent shook his head. Not typical. Something was different, if not wrong. They had to follow it, and they would follow it – but with caution.

*   *   *

"We can't carry out both operations so close together – it will look way too suspicious."

"The suspicion will be greater if they are not killed at the same time. If some are left alive to ponder the others' deaths, all kinds of red flags will be thrown up by the ones who live longer. I see your point, but we have no choice. This is the way Xenan ordered it, and this is the way it's going to happen. All the deaths will have independent explanations. The best way to sell those explanations will be to make sure no one's around to offer a different one."

*   *   *

"You must think I'm a fool." The inmate in Richard Serrap's city sneered at his fellow inmate. "You're in here because you're a crook like me, and you expect me to believe that you'll pay my family some extravagant amount of money after I'm dead! Likely! I'll blow myself up for you, and then they'll never see a cent! If you want to pay someone to do your dirty work, you better come up with a better story than that!"

"But I'm not a crook like you."

"You think I'm blind, as well as a fool! You've been in here for months. No one volunteers to live the life of a state prison inmate."

"How often have you seen me?"

"I don't know." The answer was short and frustrated, and the prisoner was getting tired of this guy he thought was a liar. He began to walk away. A strong arm grabbed him and turned him around. The face of the so-called non-prisoner now commanded attention.

"Think. In the last seven months, you've seen me six times at most – that's all I've been here. And I'm leaving again now. Want to come?"

"If what you're saying is true, then to come means that I die."

"But your family gets $100,000. You have no possibility of parole. You can die in here and leave your family nothing, or you can die on the outside and send your kids to college."

"$200,000 – delivered exactly the way I will demand, before my death."

"Come with me."

*   *   *

Linda Frederick dressed carefully. She had anticipated this meeting for more than a week. Her recent shopping spree had been well-timed. Dark blue dress, diamond and sapphire necklace, matching earrings. A year ago, she never would have paid such an exorbitant fee for the shoes. But that was then. Yes, she liked the new direction her life was taking.

The limo delivered her to a new meeting location. Across the table, Xenan removed a manila envelope from a now-empty briefcase and handed it to her. Then he scowled. He put his hand out, saying, "Give it back." She did so. He placed the envelope back in the briefcase, and slid the briefcase over next to her chair. "Keep it." His tone implied an annoyance that she had not thought to bring her own.

Linda was incensed for a moment. It had been a dinner invitation, and it was not accompanied by an instruction to bring a briefcase. Nor had he indicated the need for one at the last meeting. It wasn't fair that he berate her for not bringing something that she had no indication of needing. She was tempted to address the issue, but decided that, with this man, she had to choose her battles very carefully. Besides, she liked the idea of keeping his briefcase. She liked it a lot.

Xenan observed the play of emotions in the woman opposite him. She had passed the test well, and he was pleased that he had been able to predict how she would react, and how she would choose to respond. No one did an important job for him unless he knew these things. He knew these things – and much more – about Linda Frederick. She would be appalled to learn of some of the things he knew about...things from her past that she had worked very hard to bury. Xenan noted the dress and jewelry she wore. He watched her enjoy his notice. It confirmed what he had concluded at their last meeting with regard to the woman's emotions. Knowing what she had paid for it, and how, told him something more important: how she weighed emotion against risk.

Linda was enjoying this game, and relishing the opportunity to play with this particular opponent. Neither had spoken a word the last few moments, but they had communicated a great deal. She had never been high on drugs, but the feeling could not be any better than when she had recognized Xenan's test, and knew she had passed it. He noticed the way she was dressed, but she was disappointed at first; his eyes did not betray the type of interest she had hoped to evoke. But then an interesting thing had happened. He read her disappointment and, somehow, she knew it – and this pleased him. It had been an eerie type of knowledge, and something about it scared her; she was unsure who or what had put it in her mind. But she dismissed the fear, because seeing the pleasure in his eyes was what she had wanted. Gaining a level of control over this man apparently had less to do with physical games and more to do with spiritual ones. So be it. She had never played on these levels, but she was more than willing to learn. It was fun – a more exhilarating power game than she had known existed.

Xenan smiled. Let her believe she might actually exert a level of power over him someday. That would make her all the easier to control.

"The folder contains a Legislative Draft. Put your name on it and get it passed."

She nodded.

*   *   *

"How's she doing?"

"A little worse. She's trying so hard to hold her own. But pitting a two-and-a-half pound body against this infection is like asking my 91-year-old grandmother to tackle an NFL quarterback."

The M.D. smiled for the first time. "Gary, I met your grandmother, remember? And as I recall, she has so much spunk and guts that if she went charging after a quarterback, they might just be so shocked that they'd hand her the ball out of sheer admiration."

Gary smiled. "Yeah, you're probably right. Thanks for the encouragement."

The friend then got a serious expression on his face. "Kayla's going to make it. Keep spending time with her. Don't give up."

Gary nodded his thanks as his colleague turned and left. Then he went to Kayla's bedside and spent 30 minutes following his fellow doctor's advice.

*   *   *

Richard looked at Susan sitting in the recliner in their living room. Gratitude almost made his heart burst. She had improved a little. Instead of measuring the remainder of her life in days, it now looked like she had weeks again.

"Sue, how about if I bring some of our things down and put them in the guest room? The bathroom's not as big, but I want you to be able to sleep downstairs."

Susan smiled. "I think that would be wise." She tried valiantly not to be discouraged by the concession. She knew this day would come. It *would* be a lot easier to have the bedroom on the first floor. Still, the reality tempted her heart to despair: she might die without ever seeing their bedroom again. Strange that such a little thing could carry so much disappointment. "Lord, please help me. It's such a little thing, but I feel so weak that the weight of it is too much to bear. I need your help." A few tears spilled from her eyes as she looked up to see Richard coming in from the garage. He was carrying a shiny gift bag with handles. It had exactly the kind of surface she loved – silver, but it made rainbows when the light hit it at the right angle.

Richard had a pleased expression on his face as he set it down in front of her. "I knew it would be hard for you to switch bedrooms, so I got a little something to make the transition easier."

The bag weighed almost nothing, and when she looked inside, Susan found a little note that read, "My darling Sue, several days ago, I put up a bird feeder and a fancy hanging birdbath right next to the window of the guest room. I've kept them filled and clean, and a few finches have now found them. During the day, when you have to rest in bed, remember, 'Not one of them will fall to the ground apart from your Father. ...Therefore do not fear; you are of more value than many sparrows' (Matthew 10:29, 31). All my love, Rich."

There was no way she was going to stop the tears, but that was OK. Richard knew they were happy tears.

They were also thankful tears. Later, when she had the strength, she would have to record this event in her prayer journal. Seldom had God answered a prayer so very quickly – within seconds of the words leaving her heart – and in so special a way. The birds would make the move to the new bedroom so much easier, even enjoyable.

18

Having traded their prison garb for street clothes, the two men now sat in a Cadillac with a third man. The prisoner who had been serving a life sentence without the possibility of parole broke the silence after the fifth mile. He was looking around, relishing sights that he thought never to see again. "I'm really out." Five more miles, and then he turned to the other passenger in the back seat, the man who had approached him in the prison yard. His eyes squinted, as if in deep thought, as he asked, "Who *are* you, that you could walk me out that easy?"

"Nobody that important. But my boss gets people like you out whenever he wants. You're not allowed to ask who he is."

"I don't care who he is. What I want to know is, why does he hate Detective Stevens so much that he's willing to go to all this trouble to kill him?"

"Who said he hates Detective Stevens? Besides, even if he does, I don't know the answer."

"Of course you don't." The tone was cynical.

The prisoner's seat mate pulled out a hand gun and pointed it menacingly. "James, as you well know, there's a yard full of volunteers back there for the job you just accepted. If you still want it, you will act like it, and you will perform my instructions to the best of your ability." His eyes were cold as he allowed a three second pause. "Deal?"

James feigned a quiet respect, nodded, and went back to looking out the window. He felt nearly powerless against the desire to drink in the grass, the houses, the other cars – to stare and marvel like a child in a toy store. But as he rode, he forced his mind into action. If he had to die, he needed a plan to ensure his family's wealth and safety. But maybe, just maybe, he didn't have to die...

*   *   *

"I don't like it! Something's just not right!" The FBI agent named Tomilson, whose office was 30 miles from David Jacobson's research lab, was in a colleague's office.

"Tomilson, I know it chafes, but give it up. So this one slipped past you for a while. You caught it in plenty of time to prevent them from accomplishing anything. Watch for a while, make sure you know them all, and then go clean it up."

Tomilson shook his head. He had listened to his colleague's words, but the guy was missing the point. "I'm not upset that I missed it for so long. What's bugging me is where in the world these guys came from. Abib is the best mole we have, and he should have known about something this big long ago."

"Moles lie. Moles hide things."

"Not this one. He's telling me the truth, Sam. It's like this group materialized out of the woodwork. The original tip was anonymous. After it checked out, I ran it past Abib. It made *him* concerned."

"You did *what*? Tomilson, we don't give moles information; they feed us!"

Tomilson took a deep breath with his jaw clenched. Reporting to Sam on this one was a mistake, and he was incredibly frustrated with whoever had decided that Sam ought to oversee this case. "Sam, I understand your reaction; I really do. But please, read my case history with Abib – it goes back ten years. He was never on their side; he's a plant. I would stake – I have staked! – my reputation and my life on his information. He's trustworthy. I promise."

"He *was* trustworthy."

Tomilson was emphatic. "No. I know that he will never turn on me."

"That's not what I meant. I got this this morning." Sam handed Tomilson a thin manila folder. "He's dead."

The announcement hit agent Tomilson so hard that sound and light faded for a moment. He took the file, left Sam's office, and went to his own desk to read it. While he walked, he couldn't help but wonder if he had gotten Abib killed.

* * *

Linda Frederick was seldom taken aback by a liberal proposal. But as she read the legislative draft that Xenan had handed her, her eyes got wide. It wasn't that she disagreed with it; Linda had never seen liberal legislation that she wouldn't vote for. But she did believe in slow, careful, thought-out progress. Too much, too fast, and it was possible to shoot oneself in the foot.

What she held in her hands right now was daring. Maybe too daring. Or was it? One thing she did know: she was not going to introduce this on the Floor until she had discussed it with Xenan himself. She was willing to help him make his version of reality into history; it served her desires and purposes. But she knew that she would be unable to do this until she understood better how he thought and why. Only then would she have what she needed and still lacked – his level of certainty and, therefore, his level of confidence.

To Linda's way of thinking, this was reason enough to spend more time with the man. She had to get inside his brain. He could not object, because her motive was the attainment of his goal.

What Linda did not realize was that Xenan had already anticipated this reaction on her part. He would comply, but his reasoning was more accurate that hers: he would not be allowing her inside his head; he would be placing his thoughts in her head.

* * *

"Kayla? Can you hear me?" Dr. Gary Jones sat beside the tiny patient's bed. He was whispering just loud enough for her to be able to hear. The tone of his voice was very gentle. "Kayla, you have to fight, sweetheart. I know, it's hard. It's so hard when you're tiny, and weak, and the bad virus is big and powerful. But you can do it. Don't give up. The medicine that is supposed to help you isn't helping as much as I want. But I'm not going to give up, and so you can't either. OK?"

Kayla began breathing deeper and stronger, indicating that she not only heard his voice, but liked it. Gary continued speaking in that gentle whisper. His words were no longer directed to the tiny girl, but to her Healer. As the prayer tumbled from his lips, his head knowledge clashed with his heart's desire: if the infection was not brought under control soon, the child would be in serious peril.

* * *

Susan Serrap prayed for her husband for five minutes before she realized that physical exhaustion was overcoming her body. It was so frustrating! All she was doing was lying here, in bed, and she couldn't stay awake for any length of time before sleep claimed her. She fell asleep praying.

* * *

That afternoon, Richard knocked on Ken's door. "Ken, any word on that Judge's decision?"

Ken nodded, but his expression immediately told Richard that the news was not good.

"So *now* what? Can we appeal higher?"

Ken was enigmatic. "Yes and no. For some reason – please don't ask me the legal explanation – the lawyer found a way to appeal again. But to be honest, by the time the decision comes, the case may be over."

Richard was mad; angrier than he had been in years. The temptation was to punch the wall, kick the door, lash out somehow. With great strength, he retained his self-control – barely. "How in the world are we supposed to do our jobs?!?!"

Ken understood. When he had gotten the phone call earlier, he had lost the self-control that Richard was holding on to by a shoestring. Afterwards, for what may have been the first time in his life, Ken had prayed. Not for himself. He and God weren't yet on speaking terms where his life was concerned. But Ken was a police officer who wanted to see justice done, and it galled him that a handful of liberal judges seemed intent on preventing this. So he had prayed for a solution to their dilemma. He had prayed for a way to get Richard the information that he needed. And then, when Richard had asked his rhetorical question a moment ago, as if in answer to the earlier prayer, Ken had suddenly thought of an answer.

Richard saw Ken's countenance change from dejected and frustrated to hopeful. "Richard! A thought just came to me! The warrant you already have for Frederick's information – get to the information that way! Remember the Burlington case? We'll use the same logic!"

Richard was excited by Ken's idea, and wondered why he had not thought of it first. He calmed down, smiled, and nodded at his boss. "I'm already started."

Then he headed for his desk; he had some phone calls to make.

Ken sat still, wondering for a moment. Was that really an answer to prayer? He was honest enough with the evidence in front of him to admit that most reasonable people would say it was. This prompted another question: Why would a God whom he had ignored and offended for literally decades bother to answer one of his prayers? Ken shook his head. He didn't have time to answer that question right now. He had work to do in this office. He swiveled his chair around and applied his energy to that work. But during lunch, and again that evening, the question came back.

*   *   *

David Jacobson let out a whoop and a holler that half of the second floor of the building could hear. Several people came running, afraid that something was wrong. But as soon as they saw his face, they knew differently. Looking at them, he smiled broadly and said only, "All-staff meeting in fifteen minutes."

Jacobson sent out a quick email to alert everyone of the staff meeting, but it was not necessary. The news that something good had happened spread through the building so fast that everyone was in the meeting room in less than ten minutes.

Jacobson used 12 of the 15 minutes to throw together a brief presentation of the new research results. He knew he could have just told everyone, but that wasn't his style. He had some beautiful graphs back from the lab – at least, he thought they were beautiful – that illustrated in a moment what it would take him five minutes to explain. Deftly and efficiently, he placed them in a slideshow presentation on his laptop. Then he picked up the Bible that he kept on his desk and turned to the verse that he had studied that morning. He typed it onto the final slide of the presentation. Then he closed the screen on the computer, unplugged it from the power source, and carried it down to the meeting room. He was greeted by a roomful of expectant faces. With a teasing voice, he chided them, "You all look so impatient!"

Jacobson walked to the front of the room, where he smiled. The staff, knowing that he would not make this announcement without a presentation, had already powered up the projector and laid out all the cords, so that all he had to do was plug in the computer.

"Ladies and gentlemen, you all know what was needed before we could finish Phase Four. As I'm sure you've guessed by now, the research results from the last project just came back 15 minutes ago." He took a couple of minutes to review what had been sent to the lab, and what they had hoped to find. "And not only are the lab findings good..." Jacobson hit a button and the slide with his newest favorite graph came up. "...they are almost exactly what we predicted in print several months ago. Now, not only do we have a positive ID on our mystery inhibitor; we also have a roadmap of how an artificially-induced abortion suppresses that inhibitor."

At first, the room was dead silent as everyone stared at the graph. Then one person began applauding. Moments later, everyone was clapping. Towards the back of the room, on the face of a secretary who had spent many secret hours in prayer and fasting for this project, a tear of joy and thankfulness slipped unbidden.

Jacobson raised his hand for silence. "Thank you all for your very fine work. Let's stop and pray." For the next several minutes, the researchers and the rest of the staff poured out their gratitude to the God who was guiding their work.

"I know it's still the middle of the afternoon, but something tells me that sending you back to work right now probably won't be too useful. There's no doubt that you've earned a special break. So why don't you all take the rest of the afternoon off. It will be work as usual tomorrow morning, though. This is a wonderful breakthrough, but our work's not done. There are still two more Phases to complete. And remember, as difficult as it may be, secrecy is still the best policy for now. Let's not tip our hand before the showdown." There were many nods of understanding and agreement. "See you tomorrow."

A few of the researchers clustered around Jacobson, in order to become familiar with the finer technicalities of the new results. Several people stood around talking. Morale could not have been higher.

*   *   *

Dr. Gary Jones was sitting at his office desk when his computer bleeped. He opened the file that had just arrived, which contained Kayla's newest test results. "Good girl, Kayla!" Dr. Gary's voice was so loud that some of the people near his office poked their head in.

One of them, knowing who Kayla was, asked, "So she's going to be OK?"

Dr. Gary stopped, already on his way to her bedside. With a big smile, he nodded his head, and the other staff member smiled and nodded, too.

At the bedside, the doctor spoke gently to his little patient. "You're doing very well, Kayla! I'm so proud of you! That nasty infection is finally going away!" He laid his hand on her head. The words of his heart were now inaudible. There was still much to pray about. She was not entirely out of danger. She was still very small, barely three pounds. There were still challenges to be faced, including at least one surgery. But they had come this far. For now, this was a moment to be thankful.

19

"Here's how I want the money transferred: I want you to give me $200,000 in $100 bills. Then I want 30 minutes alone with my wife. I'll tell you where later. Give me the cash, and I will give it to her."

"No way. I'll let you give her half, but the other half gets delivered after the job is done."

James had anticipated such a response, and had decided to accept it. But he was not going to reveal this. He spent several seconds pretending to consider the deal. "OK. But I'm going to come up with the where and how that you deliver it." In fact, he already had. But time was precious, and he needed to buy as much as he could.

"Tell me within the hour, or else it gets done my way."

"I need to make a phone call."

That request brought a calculated pause. "To who?"

"To a bank."

"Not your wife?"

"Not yet."

James' captor handed him a cell phone. "Put it on speaker with that button, and if you make me suspicious, you and your family are dead."

James took the phone. "Actually, I have to make two calls. I'll need information to tell me the bank's phone number." The captor nodded.

Having obtained the bank's phone number, James explained to customer service that he had forgotten his checking account number. After a few personal questions, they gave it to him. In his head, he breathed a sigh of relief that his wife had not removed his name from the account.

Then he asked if the bank charged a fee when money was wired into the account. The call ended, and he hung up.

James handed the cell phone back. "Here's the account number that I want the money wired into after I'm dead. Now, about that meeting with my wife –"

"Enough rules from you. I'll let you see your wife this afternoon. I'll take you at 1 o'clock. You'll have at least 15 minutes. Write a note to her on this paper, telling her to come with me; that I'll take her to you. Be convincing. It's the only chance you'll have to see her."

James wrote the note, dissatisfied with the fact that his plan was falling apart. This guy was demanding too much control. For a moment, he considered trying to call the whole thing off. He had agreed to work with them out of a desire to get his family some money. Now he realized the extent to which he was threatening their safety, and regretted it. Obviously, they knew who his family was, where they were, and when they were there. He didn't like this at all. He had intended to instruct his wife to take the money and run. Change their names; hide in a little town somewhere. Now he was beginning to fear that whoever these people were, there was no way to hide from them. He finished writing the note to his wife. His captor was watching, expecting James to hand it to him. But instead, he tore it in half, and then tore it again.

Reaching for his gun, the captor asked, "What's the game?"

"No game. I may be a crook, but I love my family. Their safety is now my first concern. I don't want them to live their lives running from you. The way I see it, either you'll try to take your money back, or you'll kill them just because they're connected to me. So here's my new deal. I'll do your job, exactly as you ask, for free. But you never touch my family. You never contact them. You never give them money. You never give them anything that's going to make the cops interested in them. I don't want them connected to you in any way. They'll think I died over revenge. That's what they'll tell the cops. You won't gain a thing by harming them."

The captor's eyes narrowed, trying to discern the whole story behind these words. They made sense, if the guy really cared about his family. But this was a scenario that had not been planned for nor considered – and that made him nervous. The money that was paying for this job had planned for several possible contingencies. It was obvious that they wanted everything planned down to the finest detail. But this...? Now what?

"I'll be back." The captor, leaving James in a secure room, closed the door, and made a phone call to a man whose name he did not know.

"What?"

The captor explained the situation.

"As long as Serrap, Stevens, and Smith are all in the building, it works for us."

The captor turned around, unlocked the door, and reentered the room where James sat nonchalantly. "You have a deal."

The captor left the room, re-locking the door. James' head was spinning. As a young boy, he had become aware that his hearing was slightly better than average. It wasn't anything miraculous, but as a crook he had learned that it was just good enough to learn things that most people missed. When he had used the cell phone earlier, he had made sure the handset volume was on high before he handed it back. And just now, with his ear by the door, he had heard the mystery voice's answer. James repeated the names over and over to himself, making sure that he would not forget them: Serrap, Stevens, Smith. He had been made aware of only one target in his instructions – Stevens. But there were two more, two more that he obviously was not supposed to know about if he failed and was questioned.

But if his family's protection was his first priority, he had to follow through without saying a word. Surely, to tell the police what he knew now would get his family killed.

*   *   *

"I read the legislation you gave me. There's no way to get it passed with this President."

"I know."

"Everything depends on the next Presidential election. We may have the majority now, but I doubt we'll ever have the supermajority needed to override a veto on this."

"*Much* depends on the next election, but not everything. We already have a way around the legislature and whoever is in the White House."

"So why not use the Court on this to begin with?"

"My dear Linda, are you familiar with Marxism?"

Xenan's tone was a bit deriding, and she resented it enough to act as though his question was rhetorical. If he was not going to treat her as an equal, she would not dignify his inquiry with an answer. Knowing that he had pushed her to the limit, Xenan now retreated to what sounded like a respectful tone. She had missed the irony that he was now illustrating the very same philosophical technique that he was teaching her.

"You are familiar with the picture of the hammer?" This was actually a statement. "And the hammer is always shown in the drawn-back position. Why?" He waited to see if she could – or would – answer. She did not, but she was no longer offended. She was deep in thought, probing for the answer to his question, curious about the answer.

He continued, as if speaking a clue. "The hammer is not on its way down; but on its way up."

Linda Frederick looked at her mentor. The interest in her eyes communicated that he was feeding her intellectual candy. "You're saying that the real power is not the impact, but the moments between."

He smiled and nodded, pleased. "You will introduce this legislation with the current President. It will create a storm of outrage amongst our enemies. Since you have the majority, it will pass Congress, and the outrage will peak. The President will veto it, and the culture will think that the hammer has struck and failed. But in the months to follow, while we draw the hammer back, the culture will become desensitized to our "outrageous" proposal. The next time we present it, the radicals' outrage will surface again, but the culture will pay less attention. By then, perhaps we will have the President we need, and a simple majority in Congress will secure our desire as law. If not, we will try again, and once again, in the months to follow, the mainstream public will become desensitized further. At this point, we'll have the Court pick it up and make it law. But that's only a resort after official Constitutional channels have failed. The radicals know that we've found a way to do what we want, outside the Constitution, and they scream bloody murder every time we use it. But their little remnant can't do anything about it."

"Their 'little remnant' is pushing hard for a Constitutional Amendment to give the power back to the people."

"They won't succeed."

Linda marveled. Three words, spoken with a level of confidence and seeming prophetic ability that clashed with the reality she could see. This was where she lacked.

She had to learn and understand more. She needed to possess the level of confidence that she had just observed.

* * *

Richard entered his house and was greeted by the aroma of his mother-in-law's wonderful cooking. He couldn't figure out what it was, but boy did it smell good. He went down the hall to greet Sue first, as always. She was awake. Her journal lay open on her lap, but her eyes were smiling at the goldfinches on the window feeder. Richard just stood there for a moment, watching her joyful expression, and allowing his heart to be encouraged.

"You're looking good."

"Oh, Rich, I didn't hear you get home." She turned and held out her arms. He sat on the edge of the bed, hugged and kissed her. "Thank you again for the birds. They are such a pleasure."

"You're welcome. What else can I get you?"

Sue thought for a moment. She was past the stage of automatically saying, "Nothing." It was to the point now that, when he asked, she often did need or want something. It was tough at first, because she feared that she was taking advantage of his kindness to always make a request. But Rich had convinced her that he asked the question genuinely, and it was only her pride that was inconvenienced when she made a request.

"I'm truly OK for now; Mom was just in here. But a little later, I'd love it if you would just sit here with me and watch the birds for a little while."

Richard smiled that smile that still made her heart twitter. With his clothes changed, he entered the kitchen, where Mom was sitting at the table reading. Like a little boy whose curiosity compelled him to peek inside a Christmas present, he looked inside the pot on the stove. Swedish Meatballs!

"Mom – thank you so much!" She had stood up, and he gave her a big hug.

"You're welcome. We'll start home now. You guys have a good evening."

"Dad's here?"

"Yes, he's out back."

Richard headed out the back door. His father-in-law had mowed the lawn, and was now resting beside the pool. "Thank you, Dad."

The elder man stood up. A wise, caring face met Richard's. "You're most welcome."

Sometimes, Richard felt unable to grasp the blessings that these two wonderful people provided. "Since you're both here, why don't you stay and eat with us?"

"Not tonight. Sue's having a good day, and we were blessed to be able to share the first part of it with her. Now it's your turn. But maybe we'll plan to eat with you Sunday afternoon." Richard nodded. This was often the custom, as long as Sue was feeling up to it.

With smiles and hugs good-bye, Sue's parents left. Richard sat down on the bed next to her. She leaned into him. They watched finches and talked about their day.

Susan felt her eyelids getting heavy. She regretted being so tired, but she loved falling asleep leaning against her husband.

20

Richard knocked on Ken's open office door, and when the man at the desk looked up, Richard entered.

"Ken, I'd like to send this notice to the FBI office closest to Jacobson."

Ken took the paper. It was a copy of the notice that had already been sent to the police in Jacobson's city, notifying them that the researcher's life was probably in danger. Ken looked up, as if inviting Richard to elaborate.

Richard spoke, "The police are watching Jacobson's back. But we're dealing with an international organization. It makes sense that they will use international resources, since they have them. Having the local police alone watch for an assassin probably won't be good enough."

Ken nodded. "Send it." What he didn't tell Richard was that he had had the same thought earlier. Did Richard's God arrange coincidences like this? He wondered, but wasn't about to ask.

* * *

"That's it! I knew it!"

FBI Agent Tomilson walked so briskly towards Sam's office that a couple of people stepped out of the way before they got run over. Sam was on the phone, and Tomilson waited rather impatiently at the door while the conversation ended. As soon as the phone was away from Sam's ear, Tomilson stepped forward and handed him the electronic message.

"Sam – this is the explanation behind that sleeper cell that appeared out of nowhere."

Sam read quickly through the memo. "What makes you so sure?"

"Look at the address of Jacobson's lab."

Sam skimmed the end of the memo again. The expression on his face altered dramatically as he made the connection. His voice took on a quiet, profound tone, "They're right next door."

*   *   *

Dr. Gary Jones once again sat at Kayla's bedside. He had come to talk to her about her upcoming surgery. He didn't have an exact time yet, but probably within the next couple of days. Soon, it would be time to repair her intestine and remove the ostomy bag she had been living with.

But when he arrived at the bedside, his plans changed. Rather than talk to her, or put his hand on her head, he realized that she needed to be held. Her nurses had been trying to hold her regularly, but he had not yet taken the time to do so. With a bit of help, and the maneuvering of monitor wires and IV lines, Kayla was soon being held by her primary doctor.

Gary sat there, still and quiet. He held her near his heart, knowing that she would like that. For more than an hour, the surrounding world was forgotten by the two occupants of the rocking chair; she rested, he prayed, and they were both blessed by the other's presence.

*   *   *

"No way." The voice was as firm as it could possibly be while maintaining firm self-control. Linda knew that her arch-rival would react this way, and smiled condescendingly as he continued. "You will never make all-term abortion a Federal right, let alone Federally-funded."

Linda had gained much of Xenan's confidence, and it was audible in her voice as she replied. "We'll see."

"You'll get it out of this Committee, sure. You might even get it passed on the Floor. But something this radical? I doubt it. Even so, that's where it will die. You don't think for a second that the President will sign it? And you certainly can't override a veto. Why even bother?" Now the speaker's voice betrayed a deep-thinking curiosity.

"Maybe I've been watching you too much."

Linda's opponent, as well as several others in the room, raised their eyebrows and waited for her to explain that one.

"For months, I've watched you work for what you think is right, even in the face of certain failure. Maybe it's time I do the same."

Some of the room's occupants accepted this explanation, and even admired it. Other's, including Linda's rival, knew better. He got the last word this time. "That answer assumes that truth, right, and wrong are all relative – which they are not."

* * *

Richard sat down at his desk. Susan had not awakened before he left for work this morning. It had scared him at first. She was sleeping more and more. He had almost awakened her to kiss her goodbye; he knew she'd be upset that he had not done so. Now he regretted his decision. But he couldn't undo it. He would call her in about an hour.

Richard listened to his voicemail, and opened the snail mail that he had picked up from his box. There was a letter from Judith Kline. He opened the envelope, revealing a nice little note-card. The inside read,

Dear Detective Serrap,

I wanted to write you a note and let you know that I am still praying for you daily. The last several mornings, the Lord has put it on my heart to be praying for your safety. Be careful, please. Thank you for your continued work to discover the truth about my husband's death.

Sincerely,
Judith Kline

Richard took a moment and called Mrs. Kline to thank her. The card had meant a lot to him. Then he called Susan.

*   *   *

James looked at his captor. He felt much more like a prisoner now than when he was in the State Penitentiary. Then, he was the only one behind the bars. Now, he had put his family in with him. "I want to go over this and make sure I have it right." The other man nodded, indicating that James go ahead. "Your guy is going to be dressed like a cop. He'll walk me into the station, straight to Stevens' office. Stevens will be there. If he's not alone, we sit down and wait outside. Once he's alone, we walk in. Your guy will control Stevens while I lock the office door and close the blinds. I give him a few seconds to recognize me. If he doesn't, I tell him who I am. I ask him if he knows why I'm there. I tell him that I haven't forgotten that he's the one who put me in that hole down the road, and I would rather die, and take him with me, than live there. Your guy will watch the time. When he says, I reach in my pocket and push the button."

"Very good."

"How many minutes am I going to have to stall Stevens before your guy gives me the go ahead?"

"It depends. You just do what he says."

"You have to give me some idea! Do I need to prepare an hour's speech or something?"

"I don't have to give you anything, remember? But don't go preparing any speeches. Tell Stevens to be quiet, and don't you go making any noise, either. It should be really quick."

James thought, *Just like my death.* Out loud, he said, "There's something I don't get. You gave me this story to tell, but the only guy who's going to hear it will be dead."

"Others will hear it."

"How? And what's going to make them believe it?"

"Don't worry about the 'how.' This is what's going to make them believe it. Copy it." James' captor handed him a printed sheet, a blank sheet of paper, and a pen. He read the first sheet. This was pretty clever. It was a goodbye to his wife that doubled as a suicide note. It painted the story that his captors wanted the police to believe about his actions. James copied it as instructed. Then he addressed the envelope that he was given. As he handed everything back, he had a thought.

"You forgot something."

The captor's eyebrows rose.

"I haven't acted suicidal or vengeful the last few months."

"If necessary, there are two guards on your block that will testify otherwise."

James nodded. These guys had thought of everything, it seemed. Moreover, they had the resources to obtain whatever they needed. He was toying with the idea of crossing them, now that he knew names that he wasn't supposed to. But it would be very risky; perhaps too risky.

* * *

Richard was elated. In his hands, he held the evidence for a $10 million transfer into an overseas account that had Linda Frederick as its sole owner. Even better was the name on the account from which the money had come. Dr. Richard Kline had been right. The largest international abortion provider was the money source. He *had* to see the other places their money had gone recently. A handful of judges were doing their best to hide this information but, Lord willing, the back door that Ken had thought of the day before would get them into the account. He'd know soon enough.

* * *

Xenan conferred with the man in his office. "Everything is set?"

"Yes, sir. Tomorrow afternoon, David Jacobson, his team, and all their research will perish along with Agent Tomilson in the explosion of a dirty bomb. Simultaneously, a vengeful prisoner from the State Penitentiary will kill the cop who sent him there. Serrap and everyone who is associated with the Kline case at the precinct will be incidental casualties of that murder. Judith Kline and her son will surprise a burglar when he brings her home from her eye exam. Susan Serrap will receive a lethal injection; her death will be totally consistent with the cancer that's due to kill her any day. And Amelia, Christopher Mills' young girlfriend, will be struck by a drunk driver. Nothing will tie the cases together, and everyone who knows the big picture will be dead. Frederick's first vote will be a few days later. The most important of the court cases will probably settle without Jacobson as their star witness; worse case scenario is that our judge will throw it out for lack of evidence. By the time the appeal gets to a court that could cause us trouble, your legislation will be almost – if not already – in place."

Xenan nodded and turned his attention back to the contents of his desk. He was now alone in his office. He smiled. Having made a covenant with death, his position as successor to the Superior was almost secure.

* * *

"I think Kayla's ready for her repair surgery."

"So soon?"

"Yes. Can we fit her in tomorrow?"

"An hour ago, I would have said 'no way,' but there was just a cancellation. Two o'clock?"

"Sounds great."

* * *

"Detective Serrap? This is FBI Agent Tomilson from Seattle. I need to speak to you about the Jacobson Post."

Richard quickly grabbed a blank paper and pen. "You have my undivided attention."

"I have a very suspicious sleeper cell out here that just happens to be headquartered next door to your doctor. The only thing we have been able to confirm for sure is that there are five men who have Syrian connections. One is a munitions expert. They want everyone to believe they've been here for a while, hiding well. But my best mole assured me otherwise, and now he's dead. I want to evacuate your doctor."

"When? And to where?"

"I was told to approach him tomorrow. There's a safe house in this city."

"Don't wait that long."

"I didn't say I was going to."

Richard spoke with Tomilson several more minutes. As soon as they hung up, Richard dialed David Jacobson's cell phone. He got the answering service. He left a brief, urgent message, telling Jacobson to drop everything and go home, if he was at work, and then to call him.

* * *

James had never owned a suit this expensive. Between that and the haircut, he doubted Stevens would guess who he was. His captor took him to a post office, where he dropped his copied letter to his wife in an outside box.

"Your guy who's with me at the precinct – I'm going to kill him, too." James had been waiting for what he thought was the right time to speak to his captor about this.

"Yeah."

"Is his family getting paid?"

The look James received made it clear that he wasn't allowed to ask any more questions. But now he wondered again if he could pull off a double-cross. If he was going to be alone with a fellow inmate who had been offered the same deal he had been offered, then perhaps they could work together. Between the two of them, they knew what had to be accomplished. Could they fake their own deaths, and escape with their lives? James' mind reeled with how to approach the other man, how much time they would need, and what the best plan was. If they figured out how to set off the bomb remotely, then they might be able to fool their captors into thinking they had done everything as instructed. After all, with the bomb strapped to James' body, and the other guy in the same room, there weren't going to be whole bodies to identify anyway. Their captors had arranged all the details of their deaths; the police would not be looking for them afterward as escaped convicts.

Or, should they walk into Steven's office and immediately give themselves up? Would the police trade them their freedom for the information they possessed?

Probably not. At least, James thought, not for what he alone knew. He knew who he was supposed to kill, but that was about it. His captors had been very careful to prevent him from learning who was behind these homicides. Did the other guy know anything valuable? James considered the possibility of trying to be a lone hero. Once inside Steven's office, he could probably disarm his overseer. Then he could turn himself and the other guy over to Stevens.

Turning himself in, one way or another, seemed like the right thing to do. But it also seemed too risky. To jump the fence without setting off the bomb would make him enemies with his captors. Running from the police was one thing; they played fair. He did not want to run from these guys, who knew who and where his family was.

21

Richard saw Jacobson's name on the caller ID of his cell phone, and answered it immediately. "David, where are you?"

"On my way home, like you said. Fill me in."

Richard did so.

"I've got to get my people out of there. I grabbed the most important parts of the research, but if there's a chance we're going to loose the building, I want to take more."

"No. If they're monitoring the building, anything suspicious like that will put your people in more jeopardy. I know it's risky, but I don't think you ought to have them all leave now, either. Make sure they leave work as normal tonight – and don't come back tomorrow."

"That's three hours from now."

"I know. But Tomilson hasn't determined the extent to which your building is watched or wired. A mass exodus is likely to get everyone killed."

"So how safe is it to talk to people? Send email?"

"Do you have cell phone numbers and home addresses, so that we can contact everyone after they leave this afternoon?"

"Yes."

"We'll do it that way. Email me the contact info, and I'll make sure we find everyone."

"You'll have it in about 15 minutes."

* * *

"Tomilson – Sam wants to see you."

Great. He had been two seconds from being out the last door, and he was confident that he'd be able to get to

Jacobson without Sam being the wiser. He didn't have any evidence to mistrust the man, but based on what Richard Serrap had told him, it seemed best to move Jacobson sooner than later. Sam had been the one who had told him to approach him tomorrow.

"What's up, Sam?"

"This just came in for Agent Sanders. He's gone for the day, so I want you to take it."

Tomilson took the paper. It was an arrest warrant. There went his whole afternoon. He didn't let Sam see how he really felt. "Sure. This will probably take the rest of my day, so I'll check in with you first thing tomorrow, just before I go to pick up Jacobson."

"Very good."

Once outside the building, he dialed Agent Sanders' cell phone. "Jim, where are you?"

"Sam sent me downtown. What's up?"

"Nothing I can discuss now. Listen, don't tell anyone about this call. Later today, I need to meet with you. It's really important."

"Later then."

Tomilson began the process of serving the arrest warrant, but first he called Richard Serrap.

* * *

Susan Serrap was sleeping when her mom gently awakened her. She was holding the portable phone. Susan's heart raced and she woke up quickly, for fear that something was terribly wrong. Her mother did not wake her up for phone calls these days. Her voice was a bit shaken as she looked into her mom's eyes, dreading what she might see there. "What's wrong?"

Mrs. Sheldon was quick to calm her daughter. "There's no emergency. Richard just wants to speak to you." With that, she handed the phone over, and left the room.

"Rich? Is something wrong?"

Richard heard the fear in Susan's voice. He had expected it to be there when he chose to ask her mom to wake her up. It was for this reason that he had almost called back later. But Susan would have wanted him to wake her up. "It's OK, Sue." The tone of his voice convinced her. "I asked Mom to wake you up because I knew that's what you'd want... I called to ask if you would please stop and pray for me here at work." Sue understood immediately. For more than a decade now, she and Rich had a system. At any point in the day, if one of them was having an especially hard time, they would call the other and ask for prayer. Then they would spend a minute or two in prayer together. Sometimes they would talk for longer, but usually it wasn't necessary. The effect of that little time-out, they discovered, was absolutely amazing.

But these one and two-minute phone calls were not restricted to hard moments. Sometimes, they did it just as a way of saying, "I love you." And while Sue could summarize why she was calling, Richard could not do so, if it was case-related. For this reason, they had a code of sorts. If Richard was asking for prayer for the case he was working on, he would simply add "here at work."

The last several weeks, she and Richard had continued their prayer calls, but if Richard called when she was sleeping, he simply called back later. The fact that he had asked Mom to wake her meant that something major was happening with the Kline case. "Of course I'll pray for you." And then Sue took a minute and prayed for wisdom, safety, good information, and help for Richard and everyone he worked with. Richard was always amazed that, without being told the details of the challenge before him, Sue often prayed for exactly what was needed most.

He knew he shouldn't be surprised; he knew the verse about the Holy Spirit guiding the prayers of believers when they were unsure what to say. And he wasn't surprised; just amazed.

"Thanks, Sue."

"You're welcome. And Rich? Thanks for waking me up. I love you."

"I love you, too."

In the minutes that followed the end of the phone call, Susan continued praying. She was so thankful that Rich had awakened her. It was a blessing that he still felt able to call on her for help when something big was happening at work. She couldn't do much these days. But God had worked even this for good, for she knew that her prayers were much more fervent and effective as a result.

* * *

James could not sleep. His mind raced through scenario after scenario of what might possibly happen tomorrow. One minute, he would think that he should just abide by his captors' plan. But the desire to live was strong. Family...safety...no money...running...death... Was there a right thing to do? Did it matter if there was? Faced with the imminence of his own death, other questions that he had never bothered with surfaced as well.

* * *

Agents Tomilson and Sanders got off work at about the same time, and met in the parking lot of a local mall before going home. They were not close friends, but they had known each other and worked out of the same office for several years. Each respected the other. Tomilson explained to Sanders about Jacobson and the sleeper cell, and also what had happened that afternoon. He was fishing for Sanders' reaction. He wondered if he was overreacting about what Sam had done.

"I'm glad you told me. And no, I don't think you're overreacting. Like you said, there's no evidence that demands an inquiry at this point. It's just really weird that Sam would purposely send me out of the office so that he could assign you to do something that I was supposed to do. But if something strange happens tomorrow, an inquiry might be warranted. Honestly, I wondered what was up as soon as you told me that Sam was handling the case."

"Thanks." They shook hands and parted. Sanders went home. Tomilson called his wife to tell her that something had come up at work, and he'd be about two hours late getting home. He did not do this often, and she understood.

*   *   *

David Jacobson was not a worrier. He was wise enough to pour his worries into prayers. But right now, he was very concerned. Rather than eat dinner, he fasted and prayed.

*   *   *

"Linda, you look nice."

Linda did not answer verbally as Xenan sat down at the table across from her. She was learning quickly what pleased this man: power. And wordless conversation was often more powerful than speech.

Xenan did not share his thought that she looked like a pleased cat. He knew that she had been fishing for a compliment on her appearance for weeks now. He had saved it for just the right time. He noticed the brief case beside her chair – the one he had "given" her. He had known she would treasure it, in much the same way a high school girl treated her boyfriend's jacket. She used it every day, which meant that she took it everywhere. And as the expert had promised him, security had never touched it, even though it had been wired.

"Are you ready for your first vote in a few days?"

Her voice was an odd combination of confidence and teasing, "You mean the one we're going to lose?"

Xenan smiled. She was ready.

22

"Hi, Miss Kayla." Dr. Gary Jones smiled down at the tiny girl who was wide awake. She wasn't particularly happy, because for the moment she was not allowed to eat. Gary held her preemie-sized pacifier for her. "I know you're hungry. And it's especially hard for you, because you don't understand why we haven't fed you today. But you see, we're going to fix your tummy later today. That stinky old ostomy bag – we're going to get rid of it!" He paused and looked at her. She was calmer now, half-asleep. He stood with her another half-hour. She fell asleep while he prayed.

*   *   *

David Jacobson's cell phone had barely rung once before he answered it. "Jacobson."

"David, it's Richard Serrap."

"Are they all safe?"

"Yes."

One word, and David felt as though 500 pounds had been lifted from his shoulders. "Now what?"

"Tomilson is going to take out the sleeper cell this morning. Then we'll talk about the where and when of getting your research team back together."

"I grabbed the most important stuff, but since you contacted everyone after they left work, we missed the easy opportunity to remove a few other key things. Not that I regret the loss, even if we lose the whole building – my people's lives come first. But as soon as it's safe, I'd really like to access the lab."

"Can you give me a laundry list, or do you have to be there?"

"I – or someone on the team – would have to be there."

"Could you talk someone through the gathering process over the phone?"

David paused. "I think so."

"Plan on doing that, maybe even this morning. Tomilson and I realized you would want some things, and we discussed the possibility of sending someone into the building to collect them before the raid. But he hasn't determined whether that's safe or best yet."

"Thanks, especially for taking care of my team. And Detective? Watch your back. I'm not their only liability."

"I'll do that."

* * *

Susan was enjoying the birds at the window when she heard the clock strike. She glanced at her watch. She was surprised to notice how long she had been awake, and also by the fact that she wasn't tired yet. Perhaps the Lord had given her special strength today. She recorded her thankfulness for this in her journal, wondering what His purpose was. Suddenly, all the birds at the window flew away. A hawk had probably spooked them. With nothing special to look at, she closed her eyes and began to pray for her husband, and for his work.

* * *

Richard was off the phone with Jacobson for less than five minutes before Ken came charging into his office. The smile on his face was encouraging. Ken didn't say a word, but handed Richard a stapled packet of papers. Richard took the sheaf and skimmed the first page. The bank records!

"Ken, this might represent the best idea you've ever had!"

Ken smiled. "You're right." When all the excitement and urgency died down, he'd tell Richard the story behind it. He was curious to know what Richard would think.

Richard was now reading page after page, digesting the information as fast as he could. The same bank account that had paid Linda Frederick $10 million had distributed more than $500 million over the last month. Richard was now reading the names on the accounts to which this money had gone. A few he recognized as political lobbies, but the rest were new to him.

There were several names of individuals that intrigued Richard. It might take weeks to find all of these people, and he might not have weeks. From what he could tell, David Jacobson and his team might have been killed as early as today. Richard looked up at Ken.

"This is such a huge gold mine that I think I'll need about ten guys with shovels and pick axes."

"Just tell me who you want."

"Thanks, Ken."

*   *   *

"OK, Mrs. Kline. You're all set."

"Thank you, doctor. How long will it be before my eyes are normal again?"

"By this afternoon you should be able to take off the sunglasses. Everything might still seem a little bright. If it does, that's normal. If the light bothers you, though, put the glasses back on until tonight. Some people's eyes take a little longer to get back to normal. By tomorrow morning, you'll never know that you had this test. Do you have any other questions?"

"No. Thank you again." Mrs. Kline walked out into the waiting room, where her son was waiting to drive her home.

"All set, Mom?"

"I am."

* * *

James looked in the mirror. After not sleeping at all last night, his appearance was a strange mixture of well-dressed and haggard-looking. The suit was nearly perfect. He smiled into the mirror, trying to come up with an expression that hid the weariness in his eyes and would best please his captors. Nothing about him should draw attention.

He had come up with two separate plans, and which one he went with would depend upon his "partner." He desperately wanted several minutes alone with the man, but could not rely upon the possibility. Besides this, he could not be certain that one or both of them would be wired. One wrong word overheard by their captors, and he might be killing his family as well as himself.

A man he had never seen before unlocked the door and entered the room. He was carrying a case, which he set down on the table. "Come over here." James did so. For the next ten minutes, the man worked. The suit was a little too big in the front, and the extra space was soon taken up by a jelly-type pillow. Inside the pillow, James knew there were explosives.

After the man left, James looked into the mirror. Despite the fear and apprehension he felt over what the day held, he sort of admired the work that had just been done. No one would ever guess what he was carrying. The dark shirt he was wearing completely concealed the fact that it was a pillow, and not his stomach, that created the slight bulge in the front. A wire ran from the pillow, down his side, and into his pants pocket through a hole in the pocket next to his leg. Inside the pocket, the wire was taped to a small box with a single button. Once the wire was plugged into the box, holding down the button for five seconds would trigger the device. When James learned this, he pondered whether they would be the longest or the shortest five seconds of his life.

James had asked the bomb guy when they would be leaving. He said he didn't know. Suddenly exhausted, James laid down on the bed and dozed off. He awoke when he heard someone unlocking his door, and was almost on his feet by the time they entered.

"It's time to go." James looked around the room, as if making sure he hadn't forgotten anything. Then, with a sick feeling in the pit of his stomach, he walked out the door.

* * *

Dr. Gary came over to Kayla's bedside just before she was scheduled to be taken to the OR. "You be a good girl, Kayla. Don't worry. You'll be a little uncomfortable for a few minutes, because they're going to have that tube back in your throat for the surgery. But then you'll go to sleep. I'll be here when you wake up."

* * *

Richard was meeting with two of the men that he had chosen to dig through the bank records. "Our   first priority is to find each of these individuals. Some of them may be assassins, and we don't know their time frames. Work as fast as you possibly can."

* * *

Judith Kline walked to the car with her son, who opened the door for her. Just as she was getting in, a wave of nausea swept over her. She sat down sideways in the seat, with her feet outside the car. Closing her eyes, she leaned over and laid her head in her hands.

"Mom, what's wrong?" Concern marked his voice.

"My stomach is just feeling a bit woozy. I'll be OK in a moment. The information sheet from the doctor said this happens once in a while."

Judith's son had always been compassionate. He knelt down and gently rested his hand on her knee. "There's no rush. Can I get you something? Juice?"

"That's OK. It's passing."

"Why don't I take you to our house? I don't like the idea of leaving you at home like this."

Judith was able to sit up now. "That's not necessary, really. The nausea's going away now." She turned around in the seat and buckled her seat belt. "I'm fine." She looked up at him, as if to convince him of this. Her son's expression told her that he was not entirely convinced, but he closed the car door and headed around to the driver's side.

* * *

James was escorted to the same car that had taken him from the prison. He was placed in the back seat. In the front passenger's seat was a man dressed like a police officer. This must be his "partner."

As they drove, the captor in the back seat with James instructed him to go over the plan once more. James did so. He noticed that the guy in the front seat was not prompted in a similar manner. Obviously, the intention was to keep James from knowing as little as possible. In his mind, he reviewed the names that he wasn't supposed to have heard: Serrap and Smith.

James tried to discern if the man dressed like a police officer had ever been a prison inmate. But the man sat there wordlessly, and nothing of what James could see gave him a clue. Then the man turned, and something caught his eye. Was that what he hoped? For the first time since James realized the foolishness of getting mixed up with this bunch, hope rose in his heart. And for the first time in his life, he prayed. "God, if you're real, let that be what I think it is."

James turned to his captor. "Can I take off this suit jacket? I'm sweltering. You don't want me showing up smelling like I need a bath."

The guy nodded. James squirmed in the seat and, in the process of removing his suit jacket, got a better look at what he had hoped was a tattoo. Yes! Behind the left ear of the impersonator was a tattoo of a spider. James couldn't believe his luck: this identified the man as a member of a particular local gang.

But was it luck? The night before, James had considered praying many times. But his pride wouldn't allow it. It didn't seem right that he should turn his back on God all his life, and then come crying to Him for help the day before he died. But a moment ago, when hope had pushed desperation to the surface, he *had* prayed. And although some men would dismiss the incident as coincidence, James knew that God *had* answered.

Now what? James knew from the scenery that he had at least 20 minutes before they arrived at the precinct. His mind scrambled. Of his two original plans, he now liked one much better than the other, for now his will to live was growing stronger by the mile. If God was real – and James could no longer suppress the truth that He was – he was not ready to die. And so, for the second time in his life, James prayed. This time, he made two requests. He asked God to make it so that he could carry out the plan that allowed him to live past today. He also asked God to show him how to be ready to die.

23

"What's happening?!?!" The surgeon's hands had ceased their work, and one of the monitors was going off. The anesthesiologist had seen Kayla's blood pressure begin to drop one minute before. The team in the OR had been looking for the problem, but had been unable to halt its decline. Now they switched into emergency-mode.

* * *

Judith Kline had been riding with her eyes closed when she felt the car slow down and pull over. Her son turned to her.

"Mom, unless you're ready to tackle me over this, I'd really like to take you to our house. I can tell you haven't felt well the last ten miles. Please?"

Judith Kline smiled. "Thank you."

Robert was relieved. He turned the car around and headed toward his house. "I know you don't want to be an imposition, but trust me, you won't be. In fact, before I left this morning, Nancy told me that she was making a big batch of soup, and I was supposed to come back to your house tonight with some of it for your dinner." He laughed. "So you see – you're saving me a trip by coming now. Besides, the kids will love the opportunity to nurse their Grandma."

The nausea was still uncomfortable, but Judith Kline's heart felt lighter. She did love spending time with her grandchildren, and she was thankful for the opportunity.

* * *

As James got out of the car, he was still nervous – but less so than when he had gotten into it. The fake police officer, whom he had been instructed to call Emerson – got out also. They were in the huge parking lot next to the precinct. The two men began walking purposefully toward the building.

James suddenly realized that the car would probably remain parked where it was until their job was complete. His mind worked to deal with this possibility. He had planned that they would drop them off and drive away. Now they had baby-sitters. But this might be good – if the guys stayed in the car.

Emerson led them to the main entrance of the building. On the other side of the double doors, there was a sign-in desk. James had been told this was here. Emerson nonchalantly walked over and signed in, and James followed suit. He signed his real name, thinking that this was one of the ways that his captors had planned to make sure that his identity was discovered after his death. Then, as if in a moment of forgetfulness, he slipped the pen into the shirt pocket beneath his suit jacket.

Down the hall the two men walked. Each remained within a few feet of the other, but to the casual observer, it may not have been noticed that they were together. There was a bathroom on the left. James slowed down, which caught the eye of the other man. He nodded towards the door. "I'll be back in a moment." Emerson nodded.

* * *

Susan was praying when she heard the smoke alarm go off. She waited a few seconds. It didn't stop – and that meant that it was not just a low battery. She glanced at the clock. Her Mom would be here later this morning, but right now, she was alone in the house. She grabbed the phone by the bed and called 911.

* * *

Richard's friend Charlie came running into his office, and Richard knew something was wrong.

"There's a fire call to your house!"

Richard was out the door in seconds.

* * *

"Kayla, stay with us!" The voice was commanding, yet encouraging.

A second voice spoke almost inaudibly, frustrated and desperate. "I can't find it! Where is this coming from?!?"

Dr. Gary Jones, unaware of the events unfolding in the OR, finished going over the chart in front of him. He stretched. Time for a break. But instead of taking the ten minutes and getting a cup of coffee, as he usually did, he decided to pray for Kayla instead.

* * *

Susan was half-praying, half-thinking, trying to figure out her best option. She was really weak, and she didn't think that she could get herself all the way to either the front or the back door, and then out to safety. Perhaps adrenaline would make up for what she lacked, but she didn't think it wise to count on that. Besides, she didn't know where the fire was. The risk of being overcome by smoke entered her mind.

Susan looked at the bedroom door. It was open. But there wasn't any smoke coming in. Should she spend the strength to go over there and close it?

She was going to have to try to go out the window. But could she? She was on the first floor, and the window opened, but there was a huge shrub underneath it. Climbing out of the window would put her on top of the shrubbery – which made for a tricky exit. The shrub would not support her weight, and she feared becoming stuck and entangled in it.

She glanced back at the door again, then made her decision. She got out of bed on the side closest to the window, and pulled a light-weight blanket from the bed. She felt so weak as she made her way to the window. Perhaps more than ever before, she was aware of what the cancer had done to her body. The fact that it was going to take all of her strength to get out of the house, and that she didn't have spare strength to go close the door first, made her want to cry. But quickly she pushed those emotions aside; she would not waste energy on them.

She was half-way to the window. She would open it, and try to throw the blanket onto the shrub. If she could get it sort of spread out, then she might be able to crawl off the top of the shrub.

Suddenly, a man's face appeared in the window. Susan's first thought was that a neighbor had heard the fire call and come to help, but she didn't recognize him. She knew all the neighbors. Also, he had seemed a bit startled when he saw her working her way towards the window, which was strange. And it was far too soon for the fire department to have arrived.

He must be someone who was just passing by, heard the smoke detector, and was checking to see if anyone needed help. Judging from where he was standing, Susan figured that he must have pushed his way in between the shrub and the house – something she would never have been strong enough to do.

Over the noise of the smoke detector, Susan called to the man that she assumed had come to help her. "I'm slow, but I'm coming!"

He called back, "Hurry as fast as you can! Open the window, and I'll come in. Then I'll carry you out the front door!"

* * *

When James had gone into the bathroom, it appeared that Emerson intended to wait outside. James quickly ducked into a stall. He ripped off a length of toilet tissue, took the pen from his pocket, and scribbled a note on it. Then he laid the note on top of the toilet paper dispenser.

Ripping off a second length of toilet tissue, he wrote a second note, then replaced the pen in his pocket. Coming out of the stall, he almost ran into Emerson. It was now or never; he couldn't ask for a better opportunity. He had folded the second note into a small piece and was holding it in his hand. Now he opened his hand and offered the other man the note.

Emerson took the note and read it. The whole time, he kept a careful watch on James.

* * *

Amelia glanced at the clock in the school classroom. The last class of the day always seemed to be the longest. She glanced out the window. A warm, beautiful sunshine made her look forward to her usual walk home.

* * *

Xenan sat in a yoga position in the center of his office. In his mind, this was the most important day of his life. Soon, he would be able to boast of what his covenant with death had gained him.

24

Agent Tomilson stood in front of Sam's desk. "Are you ready to go collect that sleeper cell?"

"Yes, sir."

"I think you ought to call Jacobson first, and evacuate the building, just in case. Our first priority needs to be to protect that team. If something goes wrong during the arrest, we don't want them killed."

Tomilson had spent quite a bit of time thinking about how much to tell Sam, trying to figure out where the man's loyalties lay. "I certainly agree that protecting Jacobson's team is first priority. But if these guys are assassins, and we start evacuating everybody in broad daylight, we're likely to get them killed on the spot."

"I don't think so."

"OK. I'll call to let Jacobson know the plan, and then I'll meet you at the door in ten minutes."

"Meet me at the door?"

"I'd like your help. Jacobson's team is going to want to take as much of their research as they can; that's going to slow them down. I'll need as many men as possible in the building with me, riding these people's backs in order to get them out the door within a reasonable amount of time."

Sam looked at his watch. "I'll see you at 9:28."

Tomilson nodded and headed for his office. Now what? He had expected Sam to try to back out of helping. He sat down at his desk and picked up the phone. In case Sam was watching, he had to at least pretend to call someone. He called Jacobson, so that he could say that he had done so without lying.

Just as he hung up the phone, it rang. It was agent Sanders. "Hey, Tom. I just wanted you to know about the phone call I just received. Sam asked me to meet him this morning and give him a report. Then I'm assigned to be away this afternoon."

"Thanks. That helps."

Tomilson bit his lower lip for about two seconds, a habit he had when he was thinking quickly. Then he grabbed his jacket and headed for the door. Sam was waiting for him, but didn't have his coat on.

"You might want your jacket."

"I just got a call from Sanders – I forgot that I had to finish closing his assignment from yesterday. He asked me to take his report this morning, because he has to be gone this afternoon."

"That shouldn't take you too long. I'll wait."

"You might better go ahead without me."

Tomilson nodded and walked out the door. His team was waiting in the parking lot. He had briefed them earlier. "There's been a change of plans. But right now, we have to leave this parking lot. Jeff, drive us over to Southpoint." The van pulled out, and Tomilson reached for his cell phone. He dialed Sanders' number, well aware that he might be with Sam already.

"Sanders."

Tomilson changed his voice a little. "Mr. Sanders, this is Tommy's Mole Control returning your call." The guys closest to him in the van looked sideways.

"Sam's not here yet."

Tomilson's voice returned to normal. He quickly explained what had happened, and finished with, "As soon as I get off the phone with you, I'm going to call Franklin, and find out what my new orders are. In the mean time, I'd appreciate it if you keep an eye on Sam for me."

"Watch your back."

"Will do." Tomilson hung up that call. Then he dialed Serrap. He'd call Franklin as soon as he checked a hunch.

*   *   *

"Agent Tomilson? This is Ken Smith, Detective Serrap's superior. Serrap had to leave the office for a personal emergency. I'm familiar with the Kline case."

Tomilson filled Ken in on the situation with Sam.

"What did you say the last name is?"

"Gilland."

"Just a second." Ken had pulled a copy of the bank account records and was searching through the names. "Bingo. Samuel J. Gilland. Three weeks ago, $10 million was transferred into an account with his name on it. Good work, Tomilson."

"Would you please prepare to send a copy of those records electronically? I'll get back to you – hopefully within 15 minutes – on exactly where and to whom to send them."

"They'll be waiting."

"Thanks."

*   *   *

As Tomilson and Ken finished their conversation, three floors below Ken, Emerson finished reading James' note. It said, "Hey Spiderman. My little brother was a part of the Diamond job two years ago. Still brags on it. Don't feel bad about swallowing the money hook. Now my family's on it, too. Maybe we can live. Give Stevens 2 choices: 1. He can die. 2. He can help us fake our deaths, and we'll tell all."

Emerson pretended to write in his palm. James handed him the pen. He wrote in a blank space on James' note: "I'm not wired. You are?"

James had been paying close attention to everything that was added to his person that morning. The suit had been left in his room last night, and he had gone over it thoroughly. It was going to take so long to write everything out that he decided it was a safe gamble. Emerson might be wired, and this might be a trick, but the expression on the guy's face seemed hopeful and genuine.

"I'm not wired. I was writing because I didn't know if you were."

Emerson held out his hand. It was the gesture of a normal handshake, but James recognized it as the signal for "I'm in."

The men agreed that, if anyone started to open the door, they would each duck into a stall. No one came. They had the four uninterrupted minutes that they needed to review a whole new plan.

* * *

"Susan! Everything's OK, dear!" Susan had been concentrating so hard that she jumped half out of her skin when she heard her mother's voice yell from the door. She turned around so quickly that she lost her balance and almost fell.

"MOM?" she practically yelled, in order for her voice to be audible. The smoke alarm was still making a racket, and Mrs. Sheldon walked quickly into the room, closed the door behind her, and went to her daughter's side. The smoke alarm was still loud, but with the door closed, they could at least talk to each other. "I'm so sorry! The casserole boiled over and set off the alarm. I went and got a chair to turn it off, but I still couldn't quite reach it. Then I realized that it must have awakened you, and you didn't know I was here. I just came to tell you everything's OK. Now I'll go find a neighbor who can reach it for me."

Susan pointed to the window. "He can –" but no body was there. Susan scowled. "There was a man right there a second ago." Suddenly the strain of all the exercise and the commotion got to her, and Susan's whole body swayed. Her mom quickly supported her.

"Let's get you back to bed. Then we'll worry about the smoke alarm."

It took a couple of minutes, but then Susan was lying down. "Mom, you'll hear sirens in a few minutes. I called 911." They both laughed – it was the best option at this point.

"I'll just sit with you until they get here. You look like you're about to pass out."

Susan closed her eyes. "That's how I feel."

A few minutes later, they heard the sirens. They both laughed again. Mom got up from the side of the bed. "I'll go explain, and ask one of the taller ones to turn off that noise."

* * *

A cell phone in the van rang. "Tomilson."

"It's Franklin. Take your team and wait over at the safe house with Jacobson."

"Yes, sir."

* * *

"Good work, Dr. Jeffreys."

The surgeon nodded. "Hopefully, we won't have another roller coaster like that one again. But stay sharp, everyone. There's no rule against a second complication, and this operation's not over yet."

In a different part of the hospital, unaware of the drama in the OR, Dr. Jones finished praying for Kayla and went back to work.

*   *   *

"Officer Stevens? I'm Emerson. We have an appointment."

Stevens waved Emerson into his office. Emerson and James entered, closing the door behind them. James reached for the blinds.

"What's – ?"

"Quiet!" Emerson's gun was pointed at Stevens. "Hands on top of the desk. Now, using your feet, roll your chair over in front of that file cabinet."

James walked over and stood a little closer to Stevens. "Do you recognize me?" Stevens obviously did not. But after staring a few moments, recognition washed over his face. At this point, James continued. "Listen good, and listen quick. Some guy pretending to be a prisoner at the Pen tried to hire me to kill you. See this belly? It's a bomb. The first deal was that I'd blow you up out of supposed revenge, and after we were both dead, my family'd get two hundred grand. Right now, they're not paying a cent, and I'm just trying to keep my family alive. The surest way for me to protect my wife and kids is to hit the button in my pocket and kill us both." James glanced at Emerson. "And yes, he was plannin' to die, too. Any moment now, he's supposed to get a phone call to say it's time to push the button.

"But we'd like to offer you a chance to live, and this is how: Right now, call Smith and Serrap. Say whatever you have to in order to get them into this office – and they absolutely can not tell anyone that they're coming here. If you want to live, do it NOW."

James waited. Two seconds, three seconds. He wasn't sure if the idea would work for or against them, but it was the only one he had been able to come up with last night. The way he figured, Emerson wouldn't get the phone call until someone confirmed that Smith and Serrap were in a location where the bomb was sure to kill them. What James hoped was that he could get them into this office and hide them from whoever was going to be looking for them. This would either buy the time needed to go to the second part of his plan, or else it would backfire and the phone call would come as soon as the second man walked in the door.

Stevens nodded toward his phone. James handed it to him. Stevens dialed Ken Smith's extension. "Ken, there's a matter in my office that needs your immediate attention. It's confidential, so please don't mention where you're going. Just come on down here right away, will you? And bring Richard Serrap with you."

Ken obviously answered, and Stevens hung up. "Ken's on his way, but Serrap's not in the building. There was some sort of emergency at his house."

James couldn't believe his luck! Luck? No. A few days ago, he would have written it off as luck. But not anymore. Not only did God exist, but He seemed intent on answering James' most fervent prayers. He turned to Stevens. "Just sit still. When Smith gets here, I'll explain."

* * *

Richard came around the corner and his eyes sought his house. Everything looked normal. No smoke, no fire truck, no people milling around. He pulled quickly into the driveway and walked very quickly to the door. Opening it, he called for Susan. Her mom appeared quickly from the bedroom.

"Rich, everything's OK, and I'm so sorry – we didn't know you had heard, otherwise we would have called. We must have scared you half to death."

Richard wasn't angry, just incredibly relieved. He even smiled and half laughed, to make sure his mother-in-law knew this. "What happened?"

"It's actually a little funny. Come on in. Sue will want to help tell the story."

A few minutes later, Richard and Sue's mom were in the bedroom with Susan, and the women related what had happened.

"You did exactly the right thing, Susan. Better safe than sorry. But tell me, who was the stranger at the window?"

"It's funny, Rich, I don't know. I'm pretty sure I've never seen him before in my life. And then, after Mom came in, he disappeared."

"And I was so intent on getting to Susan that I never saw him."

Richard turned to his wife. "What did he look like?"

Susan described the face of a middle-aged man, no glasses, dark hair. "It's not very specific, I know. I could only see him from the chest up, and all my effort was being spent trying to get to the window. Besides that, I only saw him for a few seconds."

Richard went over to the window, opened it, and poked his head out. "Someone *was* here. A few of the branches on the shrubbery are broken next to the house where he pushed his way through." Richard pulled his head back in and closed the window. "I'll be right back. I just want to check something outside."

Richard went out the door and around. There was no sign of anyone. There were no real footprints to be found around the shrubbery. Because he was looking for something, he thought he could see a few places where the man's foot may have been. But the evidence was so faint – even including the tiny broken branches beside the house –

that if he had not been told someone was at the window, he probably never would have known.

Something about the stranger made him nervous. Why did he disappear? Why hadn't he come to the door after he saw Sue's mom, and made sure everything was OK? Even if he had heard Mom's explanation that the casserole had boiled over, his actions still didn't make sense. If he had heard her say it was the casserole, he would have heard her explain that she couldn't reach the smoke detector to turn it off. Why didn't he offer to do that?

Something else bothered him: where had the man come from? In the neighborhood where they lived, houses were at least a quarter mile apart; it wasn't the type of area where you saw strangers walking all the time.

Richard called the precinct, and dialed Ken's extension. He got voicemail. "Hey Ken, Richard. Everything's OK. It was just Mom's cooking that set off a smoke alarm. But I'm going to stay here for a little while. Call me." Before hanging up, he dialed Charlie's extension. Charlie had probably heard on the radio that the fire was a false alarm, but he wanted to thank him for running down to his office earlier. Charlie didn't answer his phone, either.

Richard walked back in the house. Mom was in the kitchen. He went to the downstairs bedroom. Susan was resting. He quietly checked the nightstand on Susan's side of the bed. When he had moved their bedroom to this room, he had also moved Susan's pistol and pepper spray. He repositioned them closer to the front of the top drawer. He had almost left them upstairs, uncertain of Sue's ability to even use the pistol anymore. But his cautious side had won out. He was glad it had.

Richard sat down in the chair near the bed, just looking at his wife. He could see the decline in her physical body now day by day. He was losing her too fast. Was it only a few weeks ago that she had told him that she wanted to see him solve this case? And now, to a certain extent, she had: most of the major questions were answered; from this point on it was a matter of discovering all the evidence possible, charging the guilty parties, and prosecuting them. How he wanted her prayers through that whole process! But without a miracle, that wasn't going to happen. There was no way she would live that long.

He was as prepared as possible to lose her to the cancer. He was *not* willing to sacrifice even one of their few remaining days together to an assassin. Suddenly the irony of the whole situation washed afresh over him: they wanted to kill everyone who understood the truth of the deadly connection between abortion and breast cancer, even this one nearly dead woman who was perhaps living – and dying – proof. Her imminent death, plus the reality of how close he may have come to losing her even a day early, plus the exhaustion of the earlier stress, plus everything else overwhelmed him, and he lay his head in his hands and sobbed.

The noise awoke Susan. "Rich?" She held out her hand towards him, and he came to her at once, kneeling beside the bed. He continued to cry, and she stroked his hair in the most comforting manner she knew.

25

Ken Smith knocked on Officer Stevens' door. "Come in." Ken entered the office, and the first thing he saw was Stevens wave him over. He began walking that way, and then noticed the man in the nice suit over on the side of the room. He heard the door close behind him, and turned sharply to see another man, dressed like a police officer, holding a gun. Immediately, Stevens spoke. "Don't do anything, Ken. Just sit down. These men would like to speak to us."

With Ken seated near, but not too close, to Stevens, James took over the situation. "The first thing you need to know, Smith, is what Stevens knows. This belly of mine is a bomb. All I have to do is push the button in my pocket, and we're all dead. I have nothing to lose, because pushing that button is the surest way to protect my family. Understand?" Ken nodded.

"You told Stevens that Serrap is not in the building. When's he expected back?"

"I don't know."

"You must have some idea."

"There was a fire call to his house. His wife is dying of cancer. He left on the run about 30 minutes ago. I'm sure the answer to the question depends on the severity of the emergency, and I don't know that."

"Call him. Tell him whatever you have to in order to keep him away from this building. Now." James indicated that Ken use the cell phone on his waist.

Richard saw that Ken was calling him, and figured that he was returning the earlier call, as requested. He decided not to answer it at the moment. Ken wouldn't mind.

"I got his voicemail. Message?"

"Tell him to stay away."

"Richard, it's Ken. You have to stay away from the precinct building until you get an all-clear from me personally. I can't explain, but just believe me on this one, like our lives depend on it."

James nodded. "Very good. Now, I need to tell you guys a story..." Quickly and efficiently, James recounted the events that had brought himself and Emerson to the point where they were today. "Our families are our first priority. We have no idea who these guys are, and we really don't like the idea of crossing them. But we'll give you a chance to live, and we'll do what you say, if you'll protect our families – new IDs, money, the whole nine yards."

Ken nodded. "I know who they are. And if you let me make a phone call to a certain FBI agent, I think I can convince you that we're serious about trading your help for protection."

"What's the agent's name?"

"Tomilson."

"What's his phone number?"

Ken was never so thankful that the paper with Tomilson's phone number had ended up in his pocket. "It's in my pocket."

"Hand it to me." Ken did so. "OK. I'm going to check your story. Then I'm going to hand the phone to you. When I do, I want you to put it on speaker, and then I want you to do what you said – convince me that you'll trade help for protection."

James took Ken's cell phone and dialed the number.

"Agent Tomilson."

James handed Ken the phone, who quickly hit the speaker button. "Tomilson, it's Ken Smith. I have a question. If I find a couple of guys who are willing to testify against the bunch who paid Gilland, what are the odds that you'll give them and their families witness protection? They would be providing solid evidence for conspiracy to commit murder."

"I need to look over the evidence, but I'm sure we would be very interested in making sure everyone stays safe. Is this hypothetical, or do you have them in custody already?"

"It's hypothetical. I'll keep you posted."

"Need any help on this end?"

"No. That's all. Thanks."

Tomilson hung up. He didn't like the way that conversation had worked. He had just spoken with Ken ten minutes ago, to give him the instructions for emailing the financial records that he had requested. Ken should have mentioned these witnesses then. Tomilson's intuition told him that the men had been with Smith. Was there a hostage situation going on? He was more than 2,000 miles away, and did not want to overreact, or put anyone's life in danger. He dialed Richard Serrap. But he got an answering machine.

* * *

The men waiting in the parked car at the precinct made a call to someone inside the building. "What's taking so long?"

"Serrap's not in the building."

"He's supposed to be."

"Some sort of emergency at his house – a fire call. But it turned out to be a false alarm, so I'm expecting him back pretty soon."

"How are our guys doing?"

"Fine, I suppose. I walked past the office a little while ago and the door's closed and the blinds are pulled. I'm pretty sure Ken Smith is in his office, right where he's supposed to be. The door's closed. When Serrap gets back, I'll double-check."

*   *   *

Inside Stevens' office, James addressed Ken, who had just gotten off the phone with Tomilson. "Very good, Smith. Now, here's part two of my plan. I'll warn you, you probably won't like it. But I doubt that you'll change my mind. We're going to set off this bomb, in this building, and fake at least four deaths – the two of you, and the two of us. Maybe Serrap, too. It depends. I want these guys to think that Emerson and I did our jobs exactly as we were supposed to. That's the best way to make sure that our families are protected from them."

Ken shook his head and raised his hand.

"I know what you're thinkin' – once we testify, they'll know we're alive. I thought of that. But by then, our families will be safe and hidden. If the bomb doesn't go off today, our families are in danger starting tonight – and I think they've got guys inside who would get to them."

Ken nodded, thinking of Gilland. "Go on."

"So we set off the bomb, and fake the deaths of everyone they're interested in. I want you to tell me the best, safest, quickest way to get everyone else out of the building in a way that the guys waiting in the car will probably see, but won't make them suspicious."

James looked at Smith and Stevens, who looked at each other. Smith spoke to Stevens. "When was the last time we had a fire drill?"

Stevens nodded. "Good idea. But we should get Richard back first – at least, for an appearance."

James spoke up. "I was curious to see what you'd come up with. It's interesting that we both think alike. How's this: Bring Serrap back quietly. Make sure somebody who will remember sees him. Then the four of us and him have to get snuck out of this building somehow. Then set off the fire alarm – maybe start a real fire for effect. We're going to blow up this section of the building anyway. Once everyone's out, we'll hit the button. Tomorrow, when somebody takes a head count, there will be three police officers missing. I don't think you have to make up a story about the bomb, or the two of us. I think the guys who strapped this on me have the details and press releases all set. All you have to do is sit back and watch who 'finds' the answers."

"How are you going to set off the bomb remotely?" It was Stevens who spoke.

"That's your job – you have the friends who are bomb experts. I figure it doesn't have to be this exact one, as long as it looks like it was this one."

Ken looked at James. "How much time do we have?"

"I wish I knew. All we know is that someone's going to call and say it's time to push the button. Like I told you, I wasn't supposed to know yours and Serrap's names, and I'm only guessing that the phone call won't come until you're both in range. But the longer we wait, the riskier it gets. They might tell us to go ahead without Serrap."

Ken and Stevens looked at each other. Ken spoke, "I might get fired for this, but I want to try his plan. I don't want to lose a cubic inch of this building, but if they're willing to testify, the price will be worth it. I know who these guys are, how big they are, how much money's at stake. I can't think of another option that will buy us a better deal." Stevens nodded. Then they began talking as though the other two men were not even in the room.

"We need Jennings in here to look at the bomb and come up with either a remote or a double. Then I have to call Serrap and have him waiting in the wings.

"He can sign in at the front desk and go directly to the tunnel. We can be down there already."

James interrupted, "No. We can't leave this office until there's no one out there to see us."

Stevens spoke next. "What if they get the call to hit the button when the fellows in the car see everyone leave the building?"

Ken looked at James. "He'll have to wait a few minutes before pressing the button. OK?"

James looked at Emerson. "If that happens, try to stall them."

Stevens was still concerned. "It will look too suspicious – everyone pouring out of the building, except the guys that are the intended targets."

"But they don't know that I heard the names. And they never told Emerson the names. It will be a lucky coincidence."

Ken let out a doubt-filled sigh and shook his head. Then he addressed Emerson. "If they call too soon, acknowledge their order, and then tell them to listen for the grand finale. After about 30 seconds, get back on the line. Tell them James is pushing the button, but that it won't work. We can buy a few minutes 'trying to get the button to work.'"

James nodded. "Good idea. Get that Jennings guy in here. He needs to be coming to ask you a question; don't let anyone outside this office know that you asked him here."

Ken made the phone call and briefed Jennings. A few minutes later, Jennings entered the office with a briefcase of equipment. "Let's see what we have here." Jennings worked expertly with James to disarm the switch and remove the bomb from his body. As he did so, James realized that his best threat was now disabled. Emerson still had a gun, but if these cops chose to go back on their word, his plan was finished. He looked over at Ken, who understood what James was thinking.

"I won't go back on my word. He can put the gun away."

James was inclined to trust him, but did not let his expression show this. Emerson chose to keep the gun out, and James understood.

"Try Serrap again."

Ken dialed Richard's number. This time he answered. Ken briefed Richard on the plan, and hung up. "Serrap's on his way, but he won't enter the building until I say."

*   *   *

Richard hated to leave Susan, but he did so knowing that she was now well-protected. She was being moved to a different house, and three long-time, well-armed friends were watching over her.

Before leaving his house, Richard had also called someone he trusted and instructed her to quietly take Amelia into protective custody as soon as school was out.

*   *   *

While Richard parked his car and called Ken to let him know he was there, several plainclothes officers located the Cadillac that James had described. The two men were still in it, waiting. The officers' orders were to follow it once it left.

26

At the appointed time, Tomilson called Sam. "Sam, Jacobson and his team are evacuated and safe." Actually, they had never arrived at the building that morning. All had been contacted and transported to safe houses the evening before. "I'm about to pick up the sleeper cell."

There was a pause on the other end. "Sam?"

"Yeah, I'm here. Good work. I'll see you when you get back." Sam hung up the phone and hesitated. He wasn't sure what to do, and there wasn't anyone to contact and ask. He picked up his cell phone and placed a call.

"Abort the plan and get out of there."

"To where?"

"Doesn't matter. Split up. I'll call you on this number again in a few days."

"What about our money?"

"Get out!" Sam hung up. He didn't know where Tomilson was, and he was doubtful if they had time to run. Later he would find out how Tomilson had smuggled more than 50 people out of a watched building in broad daylight. Or would he? Maybe it was time to take an early retirement. Tomilson was probably going to get the cell, and they'd talk.

Sam Gilland nonchalantly picked up his jacket and closed the door to his office. He stopped at his secretary's desk to tell her he'd be back after lunch. He did not suspect that he was being tracked as he left.

* * *

Ken called Richard. Richard got out of his car and began walking towards the precinct. His mind reviewed the fact that it was possible that a bomb would go off the moment he walked in. If they knew where Ken was, and the bomb was powerful enough to kill him from the entrance, he may well be dead in a few minutes. Richard Serrap had not been afraid to die since he had become a Christian. And at this moment, he was not suicidal. But it felt strange to not be afraid at this point. Part of it, he knew, was because Susan was so close to death. For a moment, he asked himself if he wanted to live without her. He quickly shut off those thoughts. He would not go that route; it would endanger the people around him.

Richard reached the front door and greeted the woman who signed in visitors. Then he began to make his way quickly and quietly to the tunnel. Twice someone stopped him with a brief question or a hello. His cell phone rang once, but it wasn't Ken, and he didn't answer it.

Inside the tunnel, he called Ken. Then he pulled the fire alarm. He knew that this would not bring the fire department, because Ken had called in a drill. They didn't want the fire department to show up too soon, lest the firefighters be injured when they set off the bomb.

Upstairs, the five men in Stevens' office heard the alarm go off. Ken looked over at Emerson. Nodding at the gun, he commented, "Why don't you put that away now?" Emerson nodded and did so. They stood there, looking at each other, waiting. Emerson's phone still had not rung.

*   *   *

"What's going on?!?"

"It's a fire alarm. We're all leaving the building."

"Did they come out of the office?"

"No."

"Where's Smith?"

"I found out that he went down to Steven's office, so he's probably a hostage in there. Lucky, huh?"

"What about Serrap?"

"I heard he was coming back, but I haven't seen him."

"If you can, find him, leave him near the office, and make the call. Otherwise, let it go. If it's a drill, we'll continue the plan. If it's not, they might burn up instead of blow up, and we'll kill Serrap later today."

* * *

James looked out the window. It appeared that everyone had obeyed the fire alarm, which was still sounding. "It looks clear. Smith, you lead the way."

The five men stepped out of the office. Jennings had determined the extent of the bomb's power, and had rigged a remote to it. While Ken led the two convicts to the tunnel, Jennings and Stevens were to check the floors above and below to make sure they were empty. Then they were to come to the tunnel. From there, Jennings would detonate the bomb. While Smith, Stevens, Serrap, James, and Emerson disappeared, he would meld into the crowd milling around the precinct.

Ken checked his watch. Based on previous fire drills, everyone should be out by now. Only once did they see the back of someone still leaving the building. Serrap was waiting in the tunnel.

Three minutes ticked slowly by. Part way through the fourth one, Jennings and Stevens showed up, out of breath. "Let's go." The tunnel was an emergency escape. Once they reached the end of it, Jennings pressed a button, and they heard the blast.

* * *

"How'd it go?"
"We almost lost her. But she's doing fine now."

Dr. Gary Jones' face became a picture of concern. "What do you mean, you almost lost her? What happened?!?"

The surgeon explained, and Gary had a renewed appreciation for the man's skill. He did not brag at all, but his straight-forward explanation of what had happened was understood very well by the other doctor. Gary realized that the man before him had saved Kayla's life after an unexpected complication. His voice was filled with gratitude. "Thank you."

The surgeon nodded, and Gary went quickly to the bedside of his tiny patient. She was still sound asleep. He placed a hand gently on her head and silently thanked the Lord for her life.

* * *

Xenan received a phone call that troubled him greatly. He quickly hung up, and then dialed another number.

* * *

Outside the precinct, Jennings slipped unobtrusively into the commotion of the surrounding crowd. He became one of the people who had left the building when they first heard the fire alarm, and supposedly now had no idea of an explanation for the explosion that had just rocked the building. He glanced at his watch. He had placed a call to the fire department right after detonating the bomb. He could honestly say that he was one of the employees who had been evacuated after an apparent drill, but just now there had been a huge explosion. Jennings knew that the trucks would be here soon.

A brown Cadillac pulled away, followed by three different cars. Inside, one of the Cadillac's occupants made a phone call. "Serrap is the only one who may have escaped. It didn't go exactly according to the plan, because there was some sort of fire drill. Our guy never called Emerson, but the bomb went off close enough to the other two that they're definitely dead. We think the prisoners might have panicked when everyone left the building. We'll find out about Serrap this afternoon and, if he's not already dead, take care of him."

27

Linda Frederick had never seen Xenan upset. He was hiding it well now, but she had learned how to read him well enough to figure out that something was wrong. When he spoke, he sounded almost worried. "A great deal is riding on your abilities now."

Linda, unaware of the events of the previous day, thought to herself, *A great deal has always been my responsibility; nothing has changed.*

"I warn you, you may have an uphill battle. But I know we can still win." For the first time since she had met this man, the certainty in his voice was lacking. Oh, he was still confident (or was it self-confident?) on the outside. But something had changed. She wanted to know what it was. It was more than just curiosity; she needed to know, she thought, in order to do her job to the best of her ability. If the enemy had gained some sort of advantage, she wanted to know what it was. But he did not give any indication of telling her any more.

Linda's personal motives added strength to her professional ones, and she decided to take a chance. She reached across the table, took his hand, and looked at him. She was not a heartless woman, and part of her concern was genuine. "What's wrong?"

Her ignorance was enabling her to still operate with an increased level of confidence, and Xenan was reluctant to change this. He had known that she would want to know more. She was one of very few women who had ever tried to get close to him; most were too afraid. She was attractive, and for a moment he was inclined to respond to her request.

Linda glimpsed the struggle in his face. He was human after all.

The struggle didn't last long. Power and relationships did not mix. He had never had a relationship with anyone that proved worthwhile in the long run. Becoming the Superior would place him on the top of the power pyramid for the rest of his life. That had been his goal for years, and he wasn't going to change it just because of this set-back. Years ago, he had been taught that only weak people change their goals in the midst of a set-back. At this point, he never stopped to consider the extent to which what he had learned was true.

Xenan angrily pulled his hand away. "It's better if you don't know."

* * *

In a crowded church, Susan's parents, along with their grandson, Jake, sat in the front row. Hundreds of people, it seemed, passed between them and the two closed coffins. Many stopped to speak a word of encouragement and condolence.

"I'm sorry." The countenance marked by sorrow was genuine. Mrs. Sheldon returned the embrace.

"We're doing OK, really. We know where they are, and we're so thankful that there is no suffering in Heaven."

The guest repeated something that was said many times that day. "It's so tragic that he should have to die in that explosion. But they're together."

Mr. Sheldon nodded.

And so the day wore on. Jake and Susan's parents had known that it would be a hard day, but it was actually harder than any of them had anticipated. Only they knew that Richard and Susan were actually still alive. It had been decided that this was the best and safest way. It was necessary to fake the deaths of the men who had snuck out of the police station. This meant Richard would have to disappear into hiding, at least until the court trials were over. The bad guys might well be looking for him, since the odds of him actually having died in that explosion were a little too slim and convenient.

Since Susan's life was also in jeopardy, it was reported that Susan died in her sleep that night. She had already said all of her goodbyes, and no one was surprised by the news. This allowed her to go into hiding with Richard, and would give them her last days together.

Among friends and family, only Jake, their adopted son, and Sue's parents knew the truth. No one was surprised or suspicious when Mr. and Mrs. Sheldon announced that they had decided to take their RV and "go away" for a little while right after the service. It was arranged for the RV to go someplace without them, and then they would live in the other half of the townhouse/safe house where Richard and Sue were placed. After their daughter's death, they would eventually return with the RV. Jake, having returned from overseas for the funeral services, allowed everyone to believe that he would be returning to his post almost immediately. In fact, it was arranged for him to also spend his mother's last weeks in the townhouse/safe house.

The service that Susan and Richard had planned out months ago was beautiful. It was added to a little bit, because of Richard's empty coffin. But for the most part, it was unchanged. Mr. and Mrs. Sheldon, along with Jake, had been concerned beforehand about their acting abilities. It was going to be a challenge, they had thought. But throughout the fake double funeral, they discovered that they did not have to act much at all. Part of this was because they knew that Susan was so close to death. They appreciated the heartfelt condolences of so many, and stored them away in their hearts. The day would soon come when they would bring them back out, because they would be grieving for real.

* * *

Richard and the two would-be assassins sat in a windowless office, reviewing every detail that the convicts could remember. There was still a lot of evidence that needed to be brought in before the legal actions of this case could begin. He needed to get as much information as possible from the memories of these two men before another day passed. Their families were already collected and safe.

Once the interview was finished, he met with Ken. "How's it feel to be dead?"

Ken smiled. "I've never done so much work in my life!" Indeed, there had been many details to orchestrate in order to make the deaths of the five men in the police station appear genuine.

Stevens would get a little bit of a vacation, at least at first. Up to this point, he had never been a part of the Kline-Mills case, but now he would be assigned to help with it. As near as Richard and Ken could figure, he had been targeted in the precinct homicide attempts because his office was in a prime location: if Ken and Richard had both been in their respective offices, the bomb detonated from Stevens' office would have killed both Ken and Richard. Furthermore, by using Stevens' office and the scenario involving James, investigators would probably have written off the explosion as revenge, with Stevens being the only primary target; the many other deaths, including Ken's and Richard's, would have been viewed as incidental.

Richard – to the extent Susan's health allowed on a daily basis – had much work ahead of him. Ken, relieved for the moment of overseeing the other cases in his department, would partner with Richard and devote all his time to the Kline case. It was going to take both of them, along with the team newly assigned to them while in hiding, to prepare everything needed for the prosecution of perhaps more than 30 different people. They already knew that they were dealing with at least two murders – Kline and Mills, seven attempted murders in their own city, a slew of attempted murders with regard to Jacobson and his team, bribery of at least one FBI agent and one Federal Representative, and who knew what else? There were literally hundreds of millions of dollars that had to be tracked.

"Did you hear what Stevens said about being dead?" Ken was smiling.

"No."

"He said it almost got him killed. Apparently someone told his wife that he was dead before he was able to contact her. When she learned that he was OK, she was truly thankful, but she was quite upset at him for putting her through the wringer." Ken chuckled.

"I can understand that. Although, I'm sure that her initial reaction served a good purpose." Richard looked at Ken's face, not sure what he was seeing there. "Is something wrong, Ken?"

"No. I was just thinking seriously about your question, 'How's it feel to be dead?' I've actually been giving a lot of thought to that lately, and the past few days have really driven it home." Ken's serious tone melted, and he put on a more carefree manner. "You'll probably be pleased to know that I've been thinking a lot about praying lately." And then, as if to communicate that he didn't want the conversation to go any farther, he started to walk away.

"Ken –" Ken stopped and turned. "I am pleased. See you later." And then Richard turned and started "home." The safe house was home, now, because Susan was there. He was eager to see her. With today's interviews over, he would be able to work almost exclusively from the safe house. When Susan was awake, he would be with her, and when she slept, he would work, as he was able.

Ken had been very clear a few days ago that he wanted Richard to spend as much time with Susan as possible; nothing about the case was so important that it couldn't wait. Richard appreciated this. He also knew that the faster they worked, the better. Susan would come first, definitely. But as much as possible, he needed to continue the investigation.

Richard smiled and shook his head as he got in the car. Once they had been settled at the safe house, one of the first things Sue had said was, "You're going to continue the investigation from here, right?" She was really serious about wanting to see him finish this case. They had decided together that, yes, he would continue working from the safe house. But they had also come to an agreement on how they would best take advantage of this opportunity to spend more time together.

Richard's security escort began the drive to the safe house. In the back seat, Richard thought about, for the seemingly millionth time, how much he loved his wife, and how much he was going to miss her.

* * *

David Jacobson was eager to meet with his team. After several days in a safe house, separated from them and his work, he was chafing to get back to his research. It wasn't that he was ungrateful; he was very thankful that they were all alive. It was just that there was still much work that he wanted to get done before the inevitable trials.

He had spent the days in the safe house doing internet research on the lawsuits that had already been filed against abortion providers. In most cases, the plaintiffs were women who had already been diagnosed with breast cancer. Their lawyers had availed themselves of previous research which posed a connection between the fatal choice and the fatal disease, and were attempting to make the case that their clients had not been made aware of the health risk they were taking. Thus, the patients were claiming that they had been denied the right of informed consent. Many of the cases pulled out research showing the devastating long-term emotional consequences of abortion, and claimed that the providers had suppressed disclosure of this risk as well. A few were lining up witnesses to prove that just about every single third trimester abortion – and some in the second trimester – killed a baby that would survive if delivered in a hospital. Attached to the articles, he found more pictures of unborn babies – intended to underline that these children looked just like the human babies that they were – than he anticipated.

The more Jacobson read, the more his heart grieved. The truth was that all these cases were being built on solid research that existed before his breakthrough was ever publicized – and yet, everyone was acting as though it was all "new." It wasn't new at all. But because of the splash that his breakthrough had made, suddenly people were noticing things that had been ignored for years. He was thankful that, finally, some people were waking up to what had been there all along. But were there enough such people?

David Jacobson knew that the day was fast approaching when the philosophy that had triumphed in Roe v. Wade would stand trial. When that day came, he wanted his team's research mature enough to endure the scrutiny of the court – not because it was necessary to win the case, but because its lack of maturity might be used as an unconstitutional loophole. The evidence had always demanded that abortion killed one life and maimed others, especially that of the mother. For this reason, the 1973 decision had been unjust. But for decades, the public had been too indifferent to force their legislators to reign in the renegade Court. Jacobson's research would force the Court to reconsider the abortion issue. But the enemy was smart. He would try to focus the case solely on the issue of health risk to the woman. There was plenty of old evidence to substantiate this risk as very severe, but if their new research was immature, the Court might claim insufficient evidence to overturn their previous decision. The public had swallowed this lie for years, and if they were convinced that the big headlines a few months ago had been premature and unviable, then the Court might well get away with their revisionism.

As Jacobson learned about the court cases that had been filed, he pinpointed the one that, in his estimation, had the best chance of carrying the standard of truth to the public. It was not only an informed-consent case. In addition to addressing the grave health risks to the mother, it built on the testimony of one of the world's most respected human geneticists. The geneticist himself was deceased, but this case that interested Jacobson made brilliant use of testimony that the man had given in a previous Tennessee court case. The man was a genius, and yet his testimony was laid out in such a way that any layman could easily understand his explanation of why life undoubtedly begins at conception. The plaintiffs of this case intended to show that abortion destroys life on a multi-generational level. This was the case that Jacobson wanted to see decided in Washington, D.C. This case would show America the children she had chosen not to see up until this point – if she would open her eyes.

28

Xenan sat across the table from the Superior. It was the middle of the day, but the window near the table admitted little light. The dark gray clouds overhead stifled the light and created a depressing atmosphere.

"You failed."

Xenan was silent, as he was supposed to be. But it took a great deal of willpower to suppress the unearthly anger and resentment that spilled from his heart.

"You know, you're not the only overlord in the U.S. who needs this little remnant silenced." The Superior took a sip from the glass in front of him. His steely eyes read Xenan like a book. "Perhaps it's time one of them was given the responsibility of solving this little problem."

Xenan's jaw clenched so tightly it hurt. He had anticipated such a comment, but still a wrathful reaction arose within him. And then, as quickly as the intense emotion had come, it left. He had control again; he needed it to prevent a demotion. He must convince the Superior that he could complete the necessary task.

"Perhaps Gayland could succeed where you couldn't?" The Director's tone was just slightly mocking.

"Gayland does not have the financial resources I do. Besides, he's too emotion-driven."

"You're talking about an overlord nearly as powerful as yourself."

Xenan nodded coolly. The Superior was going to let him have another chance. He was certain of it now, because of the word "nearly." He sensed the favoritism that the Superior had bestowed upon him the last several years; it had not been reneged.

No more words were spoken. The Superior nodded once, and left.

Xenan smiled, but there was no one around to notice how predatory he looked.

* * *

Linda Frederick sat across the table from a different man. The old Xenan was back, more cold and confident than ever. For the first time in her life, the man scared her. He sensed this, and used it. "You haven't been very useful lately, Ms. Frederick."

Linda was not about to let even this man intimidate her. She raised her chin a fraction of an inch, and then her eyebrows, as if challenging him. "You were the one who said that our most valuable moments are when the hammer is being drawn back."

"I don't see the hammer moving." His tone was accusing.

"Maybe you can't see everything." Her tone was caustic.

"Don't lie to me, Linda."

29

Two Months Later

"Do you solemnly swear to tell the truth, the whole truth, and nothing but the truth, so help you God?"

"I do."

"You may be seated."

David Jacobson took his place in the witness stand.

*   *   *

"Hi, Miss Kayla! Look at you in your big-girl's bed!" Dr. Jones referred to the fact that, a few weeks ago, Kayla had graduated out of her incubator to a standard crib. Putting on a hospital gown, he picked her up and sat down. He spoke to her in a soft voice.

"You're doing so well! I'm so proud of you. You're getting so big. I know, five pounds isn't 'big' to most people.   But   for   you,   it's   a   wonderful accomplishment." He paused for a moment, and his face became a little more serious. "I met your adoptive Mom and Dad today. You met them earlier, huh? I like them a lot. They'll provide you with a good home." And he really meant it. He was very thankful for the husband and wife God had provided. They were truly an answer to prayer.

"If my voice sounds a little sad, it's just because I'm going to miss you. You're a very special little girl." Dr. Gary looked at the petite face, so perfect, so beautiful to him. This wasn't his final good bye, but Kayla would be going home soon. He had to get used to not seeing her all the time. He cleared his throat of the emotion that had collected there. When he spoke, his voice was bright once again. Kayla was looking at him intently.

"Would you like to hear a story, Kayla? I just happen to have time to tell you one..." For the next 20 minutes, Dr. Gary told Kayla a story about a squirrel named Herkimer.

* * *

Six Months Later

Richard Serrap stood alone in the midst of a sea of tombstones. The sun shone, and the air was cool, but comfortable. There wasn't a cloud in the sky, and somehow this intensified the lonely feeling, in spite of the sunshine. The stone in front of him said that it had been several months now since Susan had gone home. The tears were spent, but the emptiness remained.

She had died the day after Richard had arrested Xenan. It was as though that event had marked the conclusion of her prayer responsibilities on earth. Richard smiled, remembering that last evening together. She had been so weak, but aware. Her eyes had rejoiced to hear that the authority behind Dr. Kline's murder had been taken into custody.

Richard had commented that she had played just as important a role in this case as he had. In a very humble way, she had shaken her head. Now a tear slipped unbidden from Richard's eye. He looked forward to seeing her again. When he did, he suspected that he would learn that her role in the Kline case had been just as pivotal as his.

* * *

In another section of the same graveyard stood a young girl. In front of her was the grave of Dr. Edward Kline. Looking at her, one might have thought that she wanted to cry. Reaching into her pocket, she pulled out an envelope, and placed it neatly against the tombstone. Then she took a deep breath, and walked away. Had someone opened the envelope, they would have read...

Dear Doctor Kline,

You never met me. But you helped me. Over the past several years, I've watched as several of my friends got pregnant. Each time, I thought, "Thank goodness this will never happen to me." I was thankful because, no matter what way out my friends chose, there was always pain and heartache. I learned, though, that some solutions were more painful than others. A couple of my friends had abortions. These girls seemed to have it easier at first, but then I watched regret push one of them into a horrible depression. The most socially-acceptable solution in my circle is to have the baby and be a single mom. To this day, I have to admire these friends, especially the one where the baby's dad took responsibility and they got married; at least they're trying to do the right thing and be responsible. But...I've observed, it's really hard on the family and the kids. I understand, though; the idea of giving your baby away (even if its father is a jerk) just isn't *normal*. And it's even harder after everyone knows you're pregnant. I had one friend who did give her baby up for adoption, though. And of all my friends, I think she hurt the worst at the beginning. It was like she went through the grief of having a dead baby. But the baby wasn't dead, and so she always had the option of wondering... Oh, she knew her child had a good home and all – a great home. But oh, how I watched her hurt. I told myself I never wanted to do that.

Then I got pregnant. I've never been so scared in my life. I had the advantage of having watched my friends make their decisions. I seriously considered abortion. Before I got pregnant – and even after I got pregnant, I thought it was wrong. But I still considered it. From what I'd seen in my friends' lives, it really appeared to be the easiest and least painful solution. I didn't want to be a single mom – not because I was too selfish to sacrifice for my baby, but because I didn't want him to have to deal with all the problems I was seeing in my friends' kids' lives. At first, I did not even consider giving my baby up for adoption. Of all my friends, the one that chose that route has the fewest regrets today, but it was so much pain and heartache, and innocent family and friends got hurt in the devastation. I decided to have an abortion, but I needed several weeks to save enough money. I was going to do it in secret, so my parents didn't find out.

But while I was saving the money, the newspaper headlines started talking about the connection between abortion and breast cancer. I kept saving the money. To be honest, I didn't care if I died early at this point. And then the whole story about your death came out. That was when I couldn't ignore what I said I believed any more. I had said I believed abortion was wrong. But I knew it was more than just "belief." It was true. Abortion would kill my baby, and the sin would be mine. I could justify it if I wanted. But it would still be wrong. I might hurt a lot for doing the right thing – but it would be the right thing. I kept thinking about how you had died for doing the right thing.

I couldn't kill my baby. And I still didn't want to raise him without a father. The only option left was the one I'd said I'd never do – adoption.

It was even harder than I thought it would be. So many times I started to change my mind...maybe I'll keep the baby. It was real hard on my parents, too. Just the idea of their grandchild being given away was heart-wrenching. At times, I felt worse for them than for myself. They hadn't done anything wrong, but a lot of the pain fell on them. Gossip hurts. After the baby was born, I didn't want them to see him; and yet, I wanted them to see him...so many crazy, conflicting emotions...

I did it, though. I gave him away. I chose his family, and I know he will be raised right. But before I signed the last paper, I did choose a name for his birth certificate. I gave him the first name of Edward, and the middle name of Kline – after you. You were the one God used to help me do the right thing. Thank you.

S.D.G.

www.ingramcontent.com/pod-product-compliance
Lightning Source LLC
Chambersburg PA
CBHW061302210726
48293CB00003B/1072